RARE
EARTH

Rafted ice covering the subterranean ocean of Europa (moon of planet Jupiter), a possible life habitat in the outer solar system. NASA image from the Galileo spacecraft. Courtesy of NASA.

RARE EARTH

Why Complex Life Is Uncommon in the Universe

≈≈≈≈≈≈≈

Peter D. Ward
Donald Brownlee

COPERNICUS

AN IMPRINT OF SPRINGER-VERLAG

Published in the United States by Copernicus, an imprint of Springer-Verlag New York, Inc.

Copernicus
Springer-Verlag New York, Inc.
175 Fifth Avenue
New York, NY 10010

Library of Congress Cataloging-in-Publication Data
Ward, Peter Douglas, 1949–
    Rare earth : why complex life is uncommon in the universe / Peter
Ward, Don Brownlee.
        p.   cm.
    Includes bibliographical references and index.
    ISBN 0-387-98701-0 (hardcover : alk. paper)
    1. Life on other planets.   2. Exobiology.   I. Brownlee, Don.
II. Title.
QB54.W336   1999
576.8'39—dc21   99-20532

Manufactured in the United States of America.
Printed on acid-free paper.

9 8 7 6 5 4 3 2

ISBN 0-387-98701-0   SPIN 10707256

To the memory of
Gene Shoemaker
and
Carl Sagan

Contents

Preface

This book was born during a lunchtime conversation at the University of Washington faculty club, and then it simply took off. It was stimulated by a host of discoveries suggesting to us that complex life is less pervasive in the Universe than is now commonly assumed. In our discussions, it became clear that both of us believed such life is not widespread, and we decided to write a book explaining why.

Of course, we cannot prove that the equivalent of our planet's animal life is rare elsewhere in the Universe. Proof is a rarity in science. Our arguments are *post hoc* in the sense that we have examined Earth history and then tried to arrive at generalizations from what we have seen here. We are clearly bound by what has been called the Weak Anthropic Principle—that we, as observers in the solar system, have a strong bias in identifying habitats or factors leading to our own existence. To put it another way, it is very difficult to do statistics with an N of 1. But in our defense, we have staked out a position rarely articulated but increasingly accepted by many astrobiologists. We have formulated a null hypothesis, as it were, to the clamorous contention of many scientists and media alike that life—barroom-brawling, moral-philosophizing, human-eating, lesson-giving, purple-blooded bug-eyed monsters of high and low intelligence—is out there, or that even simple worm-like animals are commonly out there. Perhaps in spite of all the unnumbered stars, we are the only animals, or at least we number among a select few. What has been called the Principle of Mediocrity—the idea that Earth is but one of a myriad of like worlds harboring advanced life—deserves a counterpoint. Hence our book.

Writing this book has been akin to running a marathon, and we want to acknowledge and thank all those who offered sustaining draughts of information as we followed our winding path. Our greatest debts of gratitude we owe to Jerry Lyons of Copernicus, who invested so much interest in the project, and to our editor, Jonathan Cobb, who fine-tuned the project on scales ranging from basic organization of the book to its numerous split infinitives.

Many scientific colleagues gave much of themselves. Joseph Kirschvink of Cal Tech read the entire manuscript and spent endless hours thrashing through various concepts with us; his knowledge and genius illuminated our murky ideas. Guillermo Gonzalez changed many of our views about planets and habitable zones. Thor Hansen of Western Washington University described to us the concept of "stopping plate tectonics." Colleagues in the Department of Geological Sciences, including Dave Montgomery, Steve Porter, Bruce Nelson, and Eric Cheney, discussed many subjects with us. Many thanks to Victor Kress of the University of Washington for reading and critiquing the plate tectonics chapter. Dr. Robert Paine of the Department of Zoology saved us from making egregious errors about diversity. Numerous astrobiologists took time to discuss aspects of the science with us, including Kevin Zahnlee of NASA Ames, who patiently explained his position—one contrary to almost everything we believed—and in so doing markedly expanded our understanding and horizons. We are grateful to Jim Kasting of Penn State University for long discussions about planets and their formations. Thanks as well to Gustav Ahrrenius from UC Scripps, Woody Sullivan (astronomy) of the University of Washington, and John Baross of the School of Oceanography at the University of Washington. Jack Sepkoski of the University of Chicago generously sent new extinction data, Andy Knoll of Harvard contributed critiques by E-mail; Sam Bowring spent an afternoon sharing his data and his thoughts on the timing of major events in Earth history; Dolf Seilacher talked with us about ediacarans and the first evolution of life; Doug Erwin lent insight into the Permo/Triassic extinction; Jim Valentine and Jere Lipps of Berkeley gave us their insights into the late Precambrian and animal evolution; David Jablonski described his views on body plan evolution. We are enormously grateful to A.A. David Raup, for discussions and archival material about extinction, and to Steve Gould for listening to and critiquing our ideas over a long Italian dinner on a rainy night in Seattle. Thanks to Tom Quinn of UW astronomy for illuminating the rates of obliquity change and to Dave Evans of Cal Tech, with whom we discussed the Precambrian glaciations. Conway Leovi talked to us about atmospheric matters. With Bob Berner of Yale, we discussed matters pertaining to the evolution of the atmosphere through time. Steve Stanley of Johns Hopkins gave us insight into the Permo/Triassic extinction. Walter Alvarez and Allesandro Montanari talked with us about the K/T extinction. Bob Pepin gave us insight into atmospheric effects.

Ross Taylor of the Australian University provided useful information to us, and Jeff Marcey and Chris McKay discussed elements of the text. Doug Lin of U.C. Santa Cruz discussed the fate of planetary systems with "Bad" Jupiters. We are grateful to Al Cameron for use of his lunar formation results.

Peter D. Ward
Don Brownlee
Seattle, Washington
August, 1999

Introduction:
The Astrobiology
Revolution and the
Rare Earth Hypothesis

On any given night, a vast array of extraterrestrial organisms frequent the television sets and movie screens of the world. From Star Wars and "Star Trek" to *The X-Files*, the message is clear: The Universe is replete with alien life forms that vary widely in body plan, intelligence, and degree of benevolence. Our society is clearly enamored of the expectation not only that there is *life* on other planets, but that incidences of *intelligent* life, including other civilizations, occur in large numbers in the Universe.

This bias toward the existence elsewhere of intelligent life stems partly from wishing (or perhaps fearing) it to be so and partly from a now-famous publication by astronomers Frank Drake and Carl Sagan, who devised an estimate (called the Drake Equation) of the number of advanced civilizations that might be present in our galaxy. This formula was based on educated guesses about the number of planets in the galaxy, the percentage of those that might harbor life, and the percentage of planets on which life not only could exist but could have advanced to exhibit culture. Using the best available estimates at the time, Drake and Sagan arrived at a startling conclusion: Intelligent life should be common and widespread throughout the galaxy. In fact, Carl Sagan estimated in 1974 that a million civilizations may exist in our Milky Way galaxy alone. Given that our galaxy is but one of hundreds of billions of galaxies in the Universe, the number of intelligent alien species would then be enormous.

The idea of a million civilizations of intelligent creatures in our galaxy is a breathtaking concept. But is it credible? The solution to the Drake Equation includes hidden assumptions that need to be examined. Most important, it assumes that once life originates on a planet, it evolves toward ever higher complexity, culminating on many planets in the development of culture. That is certainly what happened on our Earth. Life originated here about 4 billion years ago and then evolved from single-celled organisms to multicellular creatures with tissues and organs, climaxing in animals and higher plants. Is this particular history of life—one of increasing complexity to an animal grade of evolution—an inevitable result of evolution, or even a common one? Might it, in fact, be a very rare result?

In this book we will argue that not only intelligent life, but even the simplest of animal life, is exceedingly rare in our galaxy and in the Universe. We are not saying that *life* is rare—only that *animal* life is. We believe that life in the form of microbes or their equivalents is very common in the universe, perhaps more common than even Drake and Sagan envisioned. However, *complex* life—animals and higher plants—is likely to be far more rare than is commonly assumed. We combine these two predictions of the commonness of simple life and the rarity of complex life into what we will call the Rare Earth Hypothesis. In the pages ahead we explain the reasoning behind this hypothesis, show how it may be tested, and suggest what, if it is accurate, it may mean to our culture.

The search in earnest for extraterrestrial life is only beginning, but we have already entered a remarkable period of discovery, a time of excitement and dawning knowledge perhaps not seen since Europeans reached the New World in their wooden sailing ships. We too are reaching new worlds and are acquiring data at an astonishing pace. Old ideas are crumbling. New views rise and fall with each new satellite image or deep-space result. Each novel biological or paleontological discovery supports or undermines some of the myriad hypotheses concerning life in the Universe. It is an extraordinary time, and a whole new science is emerging: astrobiology, whose central focus is the condition of life in the Universe. The practitioners of this new field are young and old, and they come from diverse scientific backgrounds. Feverish urgency is readily apparent on their faces at press conferences, such as those held after the Mars Pathfinder experiments, the discovery of a Martian meteorite on the icefields of Antarctica, and the collection of new images from Jupiter's moons. In usually decorous scientific meetings, emotions boil over, reputations are made or tarnished, and hopes ride a roller coaster, for scientific paradigms are being advanced and discarded with dizzying speed. We are witnesses to a scientific revolution, and as in any revolution there will be winners and losers—both among ideas and among partisans. It is very much like the early 1950s, when DNA was discovered, or the 1960s, when the concept of plate tectonics and continental drift was defined. Both of these events prompted revolutions in science, not only leading to the complete reorganization of their immediate fields and to adjustments in many related fields, but also spilling beyond the boundaries of science to make us look at ourselves and our world in new ways. That will come to pass as well in this newest scientific revolution, the Astrobiology Revolution of the 1990s and beyond. What makes this revolution so startling is that it is happening not within a given discipline of science, such as biology in the 1950s or geology in the 1960s, but as a convergence of widely different scientific disciplines: astronomy, biology, paleontology, oceanography, microbiology, geology, and genetics, among others.

In one sense, astrobiology is the field of biology ratcheted up to encompass not just life on Earth but also life beyond Earth. It forces us to reconsider the life of our planet as but a single example of how life might work, rather than as the only example. Astrobiology requires us to break the shackles of conventional biology; it insists that we consider entire planets as ecological systems. It requires an understanding of fossil history. It makes us think in terms of long sweeps of time rather than simply the here and now. Most fundamentally, it demands an expansion of our scientific vision—in time and space.

Because it involves such disparate scientific fields, the Astrobiology Revolution is dissolving many boundaries between disciplines of science. A paleontologist's discovery of a new life form from billion-year-old rocks in Africa is of major consequence to a planetary geologist studying Mars. A submarine probing the bottom of the sea finds chemicals that affect the calculations of a planetary astronomer. A microbiologist sequencing a string of genes influences the work of an oceanographer studying the frozen oceans of Europa (one of Jupiter's moons) in the lab of a planetary geologist. The most unlikely alliances are forming, breaking down the once-formidable academic barriers that have locked science into rigid domains. New findings from diverse fields are being brought to bear on the central questions of astrobiology: How common is life in the universe? Where can it survive? Will it leave a fossil record? How complex is it? There are bouts of optimism and pessimism; E-mails fly; conferences are hastily assembled; research programs are rapidly redirected as discoveries mount. The excitement is visceral, powerful, dizzying, relentless. The practitioners are captivated by a growing belief: Life is present beyond Earth.

The great surprise of the Astrobiology Revolution is that it has arisen in part from the ashes of disappointment and scientific despair. As far back as the 1950s, with the classic Miller–Urey experiments showing that organic matter could be readily synthesized in a test tube (thus mimicking early Earth environments), scientists thought they were on the verge of discovering how life originated. Soon thereafter, amino acids were discovered in a newly fallen meteorite, showing that the ingredients of life occurred in space. Radio-telescope observations soon confirmed this, revealing the presence of organic material in interstellar clouds. It seemed that the building blocks of life permeated the cosmos. Surely life beyond Earth was a real possibility.

When the Viking I spacecraft approached Mars in 1976, there was great hope that the first extraterrestrial life—or at least signs of it—would be found (see Figure I.1). But Viking did *not* find life. In fact, it found conditions hostile

Figure I.1 *Percival Lowell's 1907 globe of Mars. Some thought that the linear features were irrigation canals built by Martians.*

to organic matter: extreme cold, toxic soil and lack of water. In many people's minds, these findings dashed all hopes that extraterrestrial life would ever be found in the solar system. This was a crushing blow to the nascent field of astrobiology.

At about this time there was another major disappointment: The first serious searches for "extrasolar" planets all yielded negative results. Although many astronomers believed that planets were probably common around other stars, this remained only abstract speculation, for searches using Earth-based telescopes gave no indication that any other planets existed outside our own solar system. By the early 1980s, little hope remained that real progress in this field would occur, for there seemed no way that we could ever detect worlds orbiting other stars.

Yet it was also at this time that a new discovery paved the way for the interdisciplinary methods now commonly used by astrobiologists. The 1980 announcement that the dinosaurs were *not* wiped out by gradual climate change (as was so long thought) but rather succumbed to the catastrophic effects of the collision of a large comet with Earth 65 million years ago, was a watershed event in science. For the first time, astronomers, geologists, and biologists had reason to talk seriously with one another about a scientific problem common to all. Investigators from these heretofore separate fields found themselves at the same table with scientific strangers—all drawn there by the same question: Could asteroids and comets cause mass extinction? Now, 20 years later, some of these same participants are engaged in a larger quest: to discover how common life is on planets beyond Earth.

The indication that there was no life on Mars and the failure to find extrasolar planets had damped the spirits of those who had begun to think of themselves as astrobiologists. But the field involves the study of life on Earth as well as in space, and it was from looking inward—examining this planet—that the sparks of hope were rekindled. Much of the revitalization of astrobiology came not from astronomical investigation but from the discovery, in the early 1980s, that life on Earth occurs in much more hostile environments than was previously thought. The discovery that some microbes live in searing temperatures and crushing pressures both deep in the sea and deep beneath the surface of our planet was an epiphany: If life survives under such conditions here, why not on—*or in*—other planets, other bodies of our solar system, or other plants and moons of far-distant stars?

Just knowing that life can stand extreme environmental conditions, however, is not enough to convince us that life is actually *there*. Not only must life be able to *live* in the harshness of a Mars, Venus, Europa, or Titan; it must also have been able either to *originate* there or to travel there. Unless it can be shown that life can form, as well as live, in extreme environments, there is little hope that even simple life is widespread in the Universe. Yet here, too, revolutionary new findings lead to optimism. Recent discoveries by geneticists have shown that the most primitive forms of life on Earth—those that we might expect to be close to the first life to have formed on our planet— are exactly those tolerant life forms that are found in extreme environments. This suggests to some biologists that life on Earth *originated* under conditions of great heat, pressure, and lack of oxygen—just the sorts of conditions found elsewhere in space. These findings give us hope that life may indeed be widely distributed, even in the harshness of other planetary systems.

The fossil record of life on our own planet is also a major source of relevant information. One of the most telling insights we have gleaned from the fossil record is that life formed on Earth about as soon as environmental conditions allowed its survival. Chemical traces in the most ancient rocks on Earth's surface give strong evidence that life was present nearly 4 billion years ago. Life thus arose here almost as soon as it theoretically could. Unless this occurred utterly by chance, the implication is that nascent life itself forms— is synthesized from nonliving matter—rather easily. Perhaps life may originate on *any* planet as soon as temperatures cool to the point where amino acids and proteins can form and adhere to one another through stable chemical bonds. Life at this level may not be rare at all.

The skies too have yielded astounding new clues to the origin and distribution of life in the Universe. In 1995 astronomers discovered the first extrasolar planets orbiting stars far from our own. Since then, a host of new planets have been discovered, and more come to light each year.

For a while, some even thought we had found the first record of extraterrestrial life. A small meteorite discovered in the frozen icefields of Antarctica appears to be one of many that originated on Mars, and at least one of these may be carrying the fossilized remains of bacteria-like organisms of extraterrestrial origin. The 1996 discovery was a bombshell. The President of the United States announced the story in the White House, and the event triggered an avalanche of new effort and resolve to find life beyond Earth. But evidence—at least from this particular meteorite—is highly controversial.

All of these discoveries suggest a similar conclusion: Earth may not be the only place in this galaxy—or even in this solar system—with life. Yet if other life is indeed present on planets or moons of our solar system, or on far-distant planets circling other stars in the Universe, what kind of life is it? What, for example, will be the frequency of *complex metazoans*, organisms with multiple cells and integrated organ systems, creatures that have some sort of behavior—organisms that we call animals? Here too a host of recent discoveries have given us a new view. Perhaps the most salient insights come, again, from Earth's fossil record.

New ways of more accurately dating evolutionary advances recognized in the Earth's fossil record, coupled with new discoveries of previously unknown fossil types, have demonstrated that the emergence of animal life on this planet took place later in time, and more suddenly, than we had suspected. These discoveries show that life, at least as seen on Earth, does not progress toward complexity in a linear fashion but does so in jumps, or as a series of thresholds. Bacteria did not give rise to animals in a steady progression. Instead, there were many fits and starts, experiments and failures. Although life may have formed nearly as soon as it could have, the formation of *animal* life was much more recent and protracted. These findings suggest that complex life is far more difficult to arrive at than evolving life itself and that it takes a much longer time period to achieve.

It has always been assumed that attaining the evolutionary grade we call animals would be the final and decisive step: that once this level of evolution was achieved, a long and continuous progression toward intelligence should occur. However, another insight of the Astrobiological Revolution has been that *attaining* the stage of animal life is one thing, but *maintaining* that level is quite something else. New evidence from the geological record has shown that once it has evolved, complex life is subject to an unending succession of planetary disasters that create what are known as mass extinction events. These rare but devastating events can reset the evolutionary timetable and destroy complex life, while sparing simpler life forms. Such discoveries again suggest that the conditions hospitable to the evolution and existence of *complex* life are far more specific than those that allow life's *formation*. On some planets, then, life might arise and animals eventually evolve—only to be quickly destroyed by a global catastrophe.

To test the Rare Earth Hypothesis—the paradox that life may be nearly everywhere but complex life almost nowhere—may ultimately require travel to the distant stars. We cannot yet journey much beyond our own planet, and the vast distances that separate us from even the nearest stars may prohibit us from ever exploring planetary systems beyond our own. Perhaps this view is pessimistic, and we will ultimately find a way to travel much faster (and thus farther), through worm holes or other unforeseen methods of interstellar travel, enabling us to explore the Milky Way and perhaps other galaxies as well.

Let's assume that we do master interstellar travel of some sort and begin the search for life on other worlds. What types of worlds will harbor not just life, but complex life equivalent to the animals of Earth? What sorts of planets or moons should we look for? Perhaps the best way to search is simply to look for planets that resemble Earth, which is so rich with life. Do we have to duplicate this planet exactly to find animal life, though? What is it about our solar system and planet that has allowed the rise of complex life and nourished it so well? Addressing this issue in the pages ahead should help us answer the other questions we have posed.

RARE PLANET?

If we cast off our bonds of subjectivity about Earth and the solar system, and try to view them from a truly "universal" perspective, we also begin to see aspects of Earth and its history in a new light. Earth has been orbiting a star with relatively constant energy output for billions of years. Although life may exist even on the harshest of planets and moons, animal life—such as that on Earth—not only needs much more benign conditions but also must have those conditions present and stable for great lengths of time. Animals as we know them require oxygen. Yet it took about 2 billion years for enough oxygen to be produced to allow all animals on Earth. Had our sun's energy output experienced too much variation during that long period of development (or even afterward), there would have been little chance of animal life evolving on this planet. On worlds that orbit stars with less consistent energy output, the rise of animal life would be far chancier. It is difficult to conceive of animal life arising on planets orbiting variable stars, or even on planets orbiting stars in double or triple stellar systems, because of the increased chances of energy fluxes sterilizing the nascent life through sudden heat or cold. And even if complex life did evolve in such planetary systems, it might be difficult for it to survive for any appreciable time.

Our planet was also of suitable size, chemical composition, and distance from the sun to enable life to thrive. An animal-inhabited planet must be a suitable distance from the star it orbits, for this characteristic governs whether the planet can maintain water in a liquid state, surely a prerequisite for animal life as we know it. Most planets are either too close or too far from their respective stars to allow liquid water to exist on the surface, and although many such planets might harbor simple life, complex animal life equivalent to that on Earth cannot long exist without liquid water.

Another factor clearly implicated in the emergence and maintenance of higher life on Earth is our relatively low asteroid or comet impact rate. The collision of asteroids and comets with a planet can cause mass extinctions, as we have noted. What controls this impact rate? The amount of material left over in a planetary system after formation of the planets influences it: The more comets and asteroids there are in planet-crossing orbits, the higher the impact rate and the greater the chance of mass extinctions due to impact. Yet this may not be the only factor. The types of planets in a system might also affect the impact rate and thus play a large and unappreciated role in the evolution and maintenance of animals. For Earth, there is evidence that the giant planet Jupiter acted as a "comet and asteroid catcher," a gravity sink sweeping the solar system of cosmic garbage that might otherwise collide with Earth. It thus reduced the rate of mass extinction events and so may be a prime reason why higher life was able to form on this planet and then maintain itself. How common are Jupiter-sized planets?

In our solar system, Earth is the only planet (other than Pluto) with a moon of such appreciable size compared to the planet it orbits, and it is the only planet with plate tectonics, which causes continental drift. As we will try to show, both of these attributes may be crucial in the rise and persistence of animal life.

Perhaps even a planet's placement in a particular region of its home galaxy plays a major role. In the star-packed interiors of galaxies, the frequency of supernovae and stellar close encounters may be high enough to preclude the long and stable conditions apparently required for the development of animal life. The outer regions of galaxies may have too low a percentage of the heavy elements necessary to build rocky planets and to fuel the radioactive warmth of planetary interiors. The comet influx rate may even be affected by the nature of the galaxy we inhabit and by our solar system's position in that galaxy. Our sun and its planets move through the Milky Way galaxy, yet our motion is largely within the plane of the galaxy as a whole, and we undergo little movement through the spiral arms. Even the mass of a particular galaxy might affect the odds of complex life evolving, for galactic size correlates with its metal content. Some galaxies, then, might be far more amenable to life's origin and evolution than others. Our star—and our solar system—are anomalous in their high metal content. Perhaps our very galaxy is unusual.

Finally, it is likely that a planet's *history*, as well as its environmental conditions, plays a part in determining which planets will see life advance to animal stages. How many planets, otherwise perfectly positioned for a history replete with animal life, have been robbed of that potential by happenstance? An asteroid impacting the planet's surface with devastating and life-exterminating consequences. Or a nearby star exploding into a cataclysmic supernova. Or an ice age brought about by a random continental configuration that eliminates animal life through a chance mass extinction. Perhaps chance plays a huge role.

Ever since Danish astronomer Nicholas Copernicus plucked it from the center of the Universe and put it in orbit around the sun, Earth has been periodically trivialized. We have gone from the center of the Universe to a small planet orbiting a small, undistinguished star in an unremarkable region of the Milky Way galaxy—a view now formalized by the so-called Principle of Mediocrity, which holds that we are not the one planet with life but one of many. Various estimates for the number of other intelligent civilizations range from none to 10 trillion.

If it is found to be correct, however, the Rare Earth Hypothesis will reverse that decentering trend. What if the Earth, with its cargo of advanced animals, is virtually unique in this quadrant of the galaxy—the most diverse planet, say, in the nearest 10,000 light-years? What if it is utterly unique: the only planet with animals in this galaxy or even in the visible Universe, a bastion of animals amid a sea of microbe-infested worlds? If that is the case, how much greater the loss the Universe sustains for each species of animal or plant driven to extinction through the careless stewardship of *Homo sapiens*?

Welcome aboard.

Dead Zones
of the Universe

Early Universe	The most distant known galaxies are too young to have enough metals for formation of Earth-size inner planets. Hazards include energetic quasar-like activity and frequent supernova explosions.
Globular clusters	Although they contain up to a million stars they are too metal-poor to have inner planets as large as Earth. Solar-mass stars have evolved to giants that are too hot for life on inner planets. Stellar encounters perturb outer planet orbits.
Elliptical galaxies	Stars are too metal-poor. Solar-mass stars have evolved into giants that are too hot for life on inner planets.
Small galaxies	Most stars are too metal-poor.
Centers of galaxies	Energetic processes impede complex life.
Edges of galaxies	Many stars are too metal-poor.
Planetary systems with "hot Jupiters"	Inward spiral of giant planets drives the inner planets into the central star.

Planetary systems with giant planets in eccentric orbits	Environments too unstable for higher life. Some planets lost to space.
Future stars	Uranium, potassium and thorium are perhaps too rare to provide sufficient heat to drive plate tectonics.

Rare Earth Factors

Right distance from star

Habitat for complex life.
Liquid water near surface.
Far enough to avoid tidal
lock.

Right mass of star

Long enough lifetime.
Not too much ultraviolet.

Stable planetary orbits

Giant planets do not
create orbital chaos.

Right planetary mass

Retain atmosphere and
ocean. Enough heat for
plate tectonics.
Solid/molten core.

Jupiter-like neighbor

Clear out comets and
asteroids. Not too close,
not too far.

A Mars

Small neighbor as
possible life source to
seed Earth-like planet,
if needed.

Plate tectonics

CO_2–silicate thermostat.
Build up land mass.
Enhance biotic diversity.
Enable magnetic field.

Ocean

Not too much.
Not too little.

Large Moon

Right distance.
Stabilizes tilt.

The right tilt

Seasons not too severe.

Giant impacts

Few giant impacts.
No global sterilizing
impacts after an initial
period.

**The right amount
of carbon**

Enough for life.
Not enough for
Runaway Greenhouse.

Atmospheric properties

Maintenance of adequate
temperature, composition
and pressure for plants
and animals.

Biological evolution

Successful evolutionary
pathway to complex
plants and animals.

Evolution of oxygen

Invention of photo-
synthesis. Not too much
or too little. Evolves at
the right time.

Right kind of galaxy

Enough heavy elements.
Not small, elliptical, or
irregular.

Right position in galaxy

Not in center, edge
or halo.

Wild Cards

Snowball Earth. Cambrian
explosion. Inertial
interchange event.

.

Why Life Might Be Widespread in the Universe

The fact that this chain of life existed in the black cold of
the deep sea and was utterly independent of sunlight—
previously thought to be the font of all Earth's life—has
startling ramifications. If life could flourish there, nurtured
by a complex chemical process based on geothermal heat,
then life could exist under similar conditions on planets far
removed from the nurturing light of our parent star, the Sun.

—Robert Ballard, *Explorations*

Several miles beneath the warm, life- and light-filled surface regions of the world's oceans lies a much harsher environment, the deep sea floor. Vast regions have little oxygen. There is no light. Much of this sea floor is composed of nutrient-poor sand, mud, or slowly precipitated manganese nodules. Temperature is a fraction above the freezing point. At least 6000 pounds of water pressure crush each square inch of matter at even average ocean basin depths. Because of these factors, except for small populations of highly specialized creatures that depend for food on the slow rain of detritus from far above, most of the deep-ocean bottom is a biological desert, long thought to be virtually lifeless and monotonous terrain.

Yet one type of environment found on the bottoms of all of Earth's oceans is neither flat nor sparsely populated. Running in linear ridges extending for thousands of miles along the sea floors are chains of active volcanic vents called deep-sea rifts. These rifts, which are situated along the margins of the great oceanic plates that make up the rocky base of the ocean floor, form undersea mountain chains. Here, in the great depth, darkness, and pressure of the sea, new crust is being created every hour, upwelling from below. These are places where the sea floor literally pulls away from itself, spreads, and in the process creates, in the endless, frigid night of the sea floor, the slow motion of tectonic plate movement known as continental drift. It seems the least hospitable environment on planet Earth. Ironically, it is teeming with life.

Amid constant earthquakes, hot magmatic lava wells up from subterranean regions in these rifts, where it encounters frigid sea water. Great gouts of this brimstone are instantly quenched as they meet the cool sea water, producing grotesque, pillow-like shapes as they turn to black rock. It is a place like no other on Earth, a region of unbelievable extremes where 2000°F lava meets 32°F water under a pressure of 400 atmospheres 2 miles beneath the sea. It is a zone of high-energy violence, where torrents of mineralized water flow like rivers out of the underworld, building great columns of metal precipitate from the hellish brew bubbling out of the Earth. Yet amid this deep-sea inferno, another and most curious phenomenon exists: submarine snow. Not the gentle snow that falls on land, but a blizzard of white material that flows out of the submarine fissures and then slowly settles onto the gnarled sea bottom. This "snow" is actually life, flocculated globs of microbes numbering in the billions and living amid the heat and poison spewing out of the vents. In utter darkness, unseen by any eye until a few humans probed the abyss in tiny, deep-diving submarines, life silently exists and thrives, creating this ethereal snowfall.

LOVERS OF THE EXTREME

The environments around the deep-ocean volcanic rifts can be described with a single word: extreme. Extreme heat, extreme cold, extreme pressure, darkness and toxic-waste waters are conditions seemingly inhospitable to every living thing. Yet over the past two decades, oceanographers and biologists who have braved the perils of the long trip to this depth in their small submarines have made stunning discoveries. The finding of bizarre tube-worms and clams was completely unexpected, but even this life is conceivable to us, for it exists in the warmed waters around the volcanic vents. What was not expected, however, was that life could live not only around, but also amid, the vents. Within these scalding cauldrons of superheated water, a rich diversity of microbial entities grow and thrive at temperatures far too hot for any animal. Yet here, indisputably, is life, in a region previously thought as sterile as Mars.

It is just such environments on Earth that may hold the most important clues to the possibility of extraterrestrial life on a place such as Mars. If the harsh hydrothermal vents can harbor life, why not the inhospitable habitats of Mars, or Europa (a moon of Jupiter), or unnumbered planets farther away as well? Life *does* exist in the hydrothermal vents of the deep sea, just as it does in other seemingly sterile habitats where organisms have recently been discovered, such as deep underground in cold basalt, in sea ice, in hot springs, and in highly acidic pools of water. Because of *where* they live, the microorganisms in these uninviting places have been dubbed *extremophiles*, "creatures that love the extreme."

The discovery that life is abundant and diverse in extreme environments is one of the most important of the Astrobiological Revolution. It gives us

3

hope that microbial life may be present and even common elsewhere in the solar system and in our galaxy, for many environments on Earth that are now known to bear extremophile life are duplicated on other planets and moons of the solar system.

The majority of research on extremophiles has centered on two types of habitats: the undersea hydrothermal vents described above and the terrestrial equivalents of the hydrothermal vents: geysers and hot pools on land. Volcanic processes create both of these habitats, and accordingly, they provide windows into the deep Earth. Life is tougher than we thought. If bacteria-like organisms can inhabit high-temperature geysers, they can live deep in Earth's crust in the subterranean blackness and heat of the underworld. The deep-ocean hydrothermal vents, and the hot springs and geysers of volcanic regions on land, are places where these previously unknown, deep-Earth assemblages of microbes can be observed and sampled. And they may also offer windows into regions where extraterrestrial life may exist on other planets and moons.

The first extremophiles were discovered not in deep-sea settings but in the geysers of Yellowstone National Park. There, in the early 1970s, microbiologist Thomas Brock and his colleagues discovered "thermophilic" extremophiles, microbes capable of tolerating temperatures in excess of 60°C, and they soon thereafter recovered microbes that could live at 80°C. Since then a variety of such extreme-heat-loving microbes have been isolated from hot springs at many localities around the world. Until that time it was believed that *no* life of any sort could live at temperatures much above 60°C, just as it is still believed that no *multicellular* organisms (such as animals or complex plants) can tolerate temperatures above 50°C. Yet, many hot springs extremophiles thrive in temperatures above 80°C, and some can live in temperatures above that of boiling water, 100°C. In contrast, the majority of bacteria grow best at 20–40°C. Discovery of these hot springs extremophiles inspired the search for similar microbes in the deep-ocean hydrothermal settings.

The deep-sea vents are characterized by three conditions previously considered deleterious to life: high pressure, high heat, and lack of light. Because of the great pressures encountered deep in the sea, water can be heated well past its boiling point at Earth's surface. The highest temperatures encountered in these environments can exceed 400°C. When this superheated, mineral-rich water hits the near-freezing sea water surrounding the vents, it is rapidly cooled, although extensive zones of water well above 80°C are found around the vents.

The submarine hydrothermal vent systems cover enormous lengths of the sea floor and may be one of the most unique habitats on Earth. However, they were virtually unknown before the 1970s because of their remoteness and depth. Since the advent of deep-diving submarines such as *Alvin*, these habitats have been intensively studied. The superheated water gushing out of the vents, once thought too hot for life, is now known to be inhabited by a diversity of microbial life, which appears to provide food for a whole host of larger organisms living around the vents. The abundant microbes thus form the base of a deep-sea food chain that requires neither light nor photosynthesisers such as plants. Most ecosystems we are familiar with have, at the base of their food chain, organisms that take carbon dioxide and light and produce living cells through photosynthesis. Light is thus the energy source that allows growth. Many of the extremophile bacteria have no need for light. They derive their energy from the breakdown of compounds such as hydrogen sulfide and methane, which fuels their metabolism. Furthermore, these organisms evolved early in earth history, and this suggests that the earliest life on our planet—and by inference on other planets as well—may be chemically fueled rather than powered by light. The implication is that light may not be a prerequisite for life.

Perhaps the most unexpected aspect of these discoveries was that many of the bacteria in these regions not only support, but also demand and thrive on, temperatures above 80°C. One species discovered in the deep-sea hydrothermal vents reproduces best in water at temperatures above 105°C and remains able to reproduce in water as hot as 112°C.

Even more startling lovers of extreme heat have recently been found in these environments. In 1993, John Baross and Jody Deming of the University of Washington published a paper entitled "Deep-sea smokers: Windows to a subsurface biosphere?" In this paper, the two oceanographers advanced the idea that the interior of Earth is home to microbes capable of living, under high pressure, at temperatures above that of boiling water—as much as 150°C. They called these organisms "super thermophilic." This bold prediction was supported when John Parkes of Bristol, England, discovered intact microbes at 169°C in a deep-sea drill core. What is the upper temperature limit for life? Microbiologists now theorize that life may be able to withstand 200°C in high-pressure environments.

Although some of them fall within the taxonomic group formally called Bacteria, the majority of these extremophilic microbes belong to the taxonomic group known as Archaea. The archaea are biological stalwarts indeed. They thrive in boiling water and live on elements toxic to other life, such as sulfur and hydrogen. The discovery of this major group of living organisms itself precipitated one of the great revolutions in biology, for their existence required a substantial reconfiguring of the time-honored model we can call the "Tree of Life," the theorized evolutionary pathway leading from the earliest life to the most complex.

THE ARCHAEANS

Biologists have long recognized that species can be grouped into hierarchical assemblages. These units are linked by lines of descent; that is, all species that make up a higher category share a common ancestor. Species are grouped into genera. (Our species is grouped, along with the extinct human forms, into the genus *Homo*. This means that all species of *Homo*, including *Homo sapiens*, *Homo erectus*, and *Homo habilis*, among others, have a common ancestor.) Genera are grouped into families, families into orders, orders into classes, classes into phyla, and phyla into kingdoms. The kingdoms have always been defined as the highest level, so they are not grouped into any higher unit. The earliest practitioners of this system, which was developed by the great Swedish naturalist Carl Linnaeus in the eighteenth century, first recognized only two kingdoms: animals and plants. As biologists invented and mastered microscopes and came to understand plants better, they increased the number of kingdoms to five: the kingdoms Animalia, Plantae, Fungi, Protozoa, and Bacteria. But the discovery of the archaea changed all of that. They are so different that they have required scientists to devise an entirely new taxonomic category of life.

The archaea have long been overlooked because they closely resemble bacteria. But once molecular biologists were able to analyze their DNA, it became clear that these tiny cells were as different from bacteria as bacteria are from the most primitive protozoans. This led University of Illinois biologist Carl Woese to propose a new category of life, the *domain*, which he placed above kingdoms. In this scheme, the five kingdoms are spread over three domains: Archaea, Bacteria, and a new category called Eucarya, which includes the plants, animals, protists, and fungi.

The domain Archaea is itself subdivided into two previously unrecognized kingdoms: the kingdom Crenarchaeota, made up of heat-loving forms, and the kingdom Euryarchaeota, which includes a few thermophiles but is composed mainly of forms that produce the organic compound methane (swamp gas) as a biological by-product of their metabolism. Most archaeans are "anaerobic"; they can live only in the absence of oxygen. This characteristic makes them prime candidates for the first life on Earth, because the newly formed Earth had no free oxygen.

Although many types of archaeans have been found in hot-water settings, it is clear that they can live in other subterranean settings, including within solid rock itself. The first clue that life might exist hundreds to thousands of meters below Earth's surface came in the 1920s, when geologist Edson Bastin of the University of Chicago began to wonder why water extracted from deep within oil fields contained hydrogen sulfide and bicarbonates. Bastin knew that both of these compounds are commonly created by bacterial life, yet the water coming from the oil wells was from environments that seemed far too deep and hot to support any sort of bacterial life discovered up to that time. Bastin enlisted the aid of microbiologist Frank Greer, and together they succeeded in culturing bacteria recovered from this deep water. Regrettably, their findings were dismissed by other scientists of the time as being due to contamination from the oil pipes, and this first interdisciplinary venture linking the fields of geology and microbiology languished, its provocative discovery ignored for more than 50 years.

The possibility that life was present deep within our planet was finally taken seriously when scientists began studying groundwater around nuclear waste dumps in the 1970s and 1980s. As ever-deeper boreholes were drilled, microbial life was routinely found at depths long thought to be too great to support life of any kind. But were the microbes found at these depths actually living there, or were they contaminants from surface regions that were picked up by the sampling equipment on its journey down? This question was not answered until 1987, when an interdisciplinary team of scientists assembled by the United States Department of Energy built a special coring device capable of drilling deep into the rock and extracting samples with no possibility of contamination. Three 1500-foot-deep boreholes were drilled at a government nuclear research laboratory near Savannah River, South Carolina. Samples brought to the surface were analyzed for microbes, and it was quickly discovered that microbial life did indeed exist at these depths and that it was rich in both number of species and number of individuals. A new habitat for life had been discovered, and the pioneering work of Bastin and Greer had been confirmed.

It is generally acknowledged that the cataloguing of Earth's species is far from complete—that many species of all groups of life, not just extremophiles, wait to be discovered. Less well known is that our understanding of the *habitats* occupied by life on this planet may be equally incomplete; the new extremophile discoveries beneath Earth's surface are proof of that. In this age of satellite surveys and global travel, it seems incongruous that there could be vast unexplored regions harboring unknown life, but this is certainly the case. Aside from Jules Verne's imaginative and prophetic novel *Journey to the Center of the Earth*, humankind has little penetrated the last frontier and the region that may hold the single largest mass of life inhabiting the planet: deep in Earth's crust.

With the discovery of deep life in South Carolina, many teams began probing ever deeper underground, trying to find the lower limit of life within the crust of Earth itself. Soon they learned that subterranean microbes could be found in most geological formations; the deep bacterial and archaean world thus appears ubiquitous under the surface. The greatest depth from which these life forms have so far been recovered is about 3.5 kilometers, at temperatures of 167°F. At such great depths, however, the population density of the microbes is low. They can live in many rock types, including both sedimentary and igneous rocks. Temperatures increase in a planet as one descends deeper into the crust. Archeans may inhabit a wide range of rock types even several miles beneath Earth's surface. Cornell University geologist Thomas Gold has gone so far as to suggest that the combined biomass of microorganisms beneath Earth's surface may be several times that of *all* organisms—great and small, complex and simple—living on the surface above. If so, microorganisms are by far the most numerous organisms on Earth!

The maximum depth at which extremophiles have been found to live is constantly being revised. In 1997 the record was 2.8 kilometers, but soon a mine located in South Africa yielded specimens from a depth of 3.5 kilometers. The basic requirements of the inhabitants of this "deep biosphere" are water; pores, in the source, of sufficient size to allow the presence of the deep microbes; and nutrients. Because the extremophiles are adapted for pressure they are virtually unaffected by the high pressures encountered at these great depths.

The nutrients used by these deep-living extremophiles come from the rocks they live in. In sedimentary rock, nutrients derive from organic material trapped at the time of the rock's deposition. The deep-biosphere microbes (the microbes living in sedimentary rock) then utilize this material for the energy and organic matter necessary for life. Oxidized forms of iron, sulfur, and manganese are also utilized as nutrients. Living in sedimentary rock thus poses no great hardship for certain archaeans and bacteria. Living in igneous rock, however, is a more difficult proposition.

Igneous rock, such as basalt (the rock that forms when lava cools and solidifies) has no (or very little) constituent organic matter. It was therefore a major surprise when scientists in Washington state discovered flourishing communities of microbes living in ancient basaltic rock in the Columbia River basin. Microbiologists Todd Stevens and James McKinley from Batelle Laboratory discovered in the 1980s that many of the bacteria they found in these rocks were manufacturing their own organic compounds, using carbon and hydrogen taken directly from hydrogen gas and carbon dioxide dissolved in the rock. They produced methane as a by-product of their synthesis, so they acquired the name methanogens. These archaea are thus autotrophs, organisms that can produce organic material from inorganic compounds. Co-occurring heterotrophic or organic-consuming microbes then ingest some of the organic material produced by these autotrophic organisms. This is (like the deep-sea vent community) an ecosystem totally independent of solar energy—independent of the surface and of light. These particular communities have been dubbed—perhaps appropriately—the SLiME communities, for "subsurface lithoautotrophic microbial ecosystem." Because their presence in these dark, sometimes hot regions of Earth's crust tells us that sunlight is not necessary to sustain life, their discovery is one of the most important ever made about the range of environments that can support life. It means that even a far-distant and relatively cold planet such as Pluto could conceivably support life in the warm, inner portions of its crust. Planets and moons far from a star may have frigid surfaces, but their interiors are warm with heat from radioactive decay and other processes.

The deep-rock microbe communities can be trapped within their host rock for millions of years. They first get into the igneous rock via flowing groundwater, but in some instances this groundwater is cut off, and yet the deep microbes persist and thrive. Samples from the Taylorsville region of Texas are thought to be 80 million years old and have grown and evolved at exceedingly slow rates. They became trapped in the hard igneous rock during the heyday of the dinosaurs and remained there, living without any contact with the rest of Earth's life, until humans released them by digging deep wells. Some of these microbes have adapted to very low levels of nutrients and tolerate extended periods of starvation.

Extremophiles are not only adapted to hot and high-pressure conditions. Other groups are found in conditions thought too *cold* for life. All animal life eventually ceases at below-freezing temperatures. When the bodies of animals are cooled below the freezing point, they can enter a state of suspended animation, but the metabolic functions do not continue. Some extremophiles, however, circumvent this. Microbiologist James Staley of the University of Washington discovered a new suite of extremophiles living in icebergs and other sea ice. This habitat was long considered too cold to harbor life, yet life has found a way to live in the ice. This particular finding is as exciting and as relevant to the astrobiologist as the heat-loving extremophiles, for many places in the solar system are locked in ice. Other extremophiles relish chemical conditions inimical to more complex life, such as highly acidic or basic environments or very salty seawater.

THE MARTIAN CONNECTION

The interest in extremophilic microbes intensified after the discovery of the now-famous Martian meteorite known as ALH 84001, a hunk of rock found in the Allan Hills region of Antarctica on December 27, 1984. After it was discovered, this piece of cosmic slag was filed away and forgotten for a decade. It was finally reexamined, however, and determined to be from Mars. A team of NASA scientists then began to probe it, and their examination culminated in the stunning announcement on August 7, 1996, that this particular piece of rock might contain fossils of Martian microbes in its stony grasp.

Of the various lines of evidence used by NASA scientists to arrive at this startling conclusion, the most fascinating were small rounded objects in the meteorite resembling fossil bacteria. And why not? Conditions on the Martian surface today are highly inimical to life: subject to harsh ultraviolet radiation, lack of water, numbing cold. The Mars Pathfinder expedition only seemed to confirm the planet's inhospitality—even for the highly tolerant extremophilic microbes. But what of the Martian *subsurface*? Perhaps life still exists in the subterranean regions of Mars, where hot hydrothermal liquid associated with volcanic centers could create small oases, a Martian equivalent of Earth's deep biosphere, replete with archaeans.

And even if life is now totally extinct on Mars, what of its past? Since the Viking landing of 1976, scientists have known that the ancient Mars had a much thicker atmosphere and had water on its surface, at least for a brief period of time. Three billion years ago, Mars could have been warmer because of its cloaking atmosphere. Such conditions still would have been too harsh for animal life, but judging from what we now know about the extremophiles on Earth, the early Martian environment would have been quite conducive to colonization by microbes. The extremophiles need water, nutrients, and a source of energy. All would have been present on Mars. It may be that life does not exist Mars today. Yet there may be a great deal that we can learn about ancient Mars in its fossil record—a fossil record perhaps populated by Martian analogs to Earth's extremophiles. Andrew Knoll of Harvard University has pointed out that for very old rocks, the fossil record may be fuller on Mars than it is on Earth, because there has been little erosion or tectonic activity on Mars to erase the billions of years of fossil records. Knoll has even told us where on Mars to search for fossils: on an ancient volcano named Apollinaris Pater, whose summit shows whitish patches interpreted to be the minerals formed by escaping gases, or in a place called Dao Vallis, a channel deposit on the flank of another ancient volcano where hot water may have flowed out from a hydrothermal system within the Martian interior. Mineral deposits there might yield a rich fossil record of ancient Martian extremophiles.

IMPLICATIONS FOR THE "HABITABLE ZONE"

The discovery of extremophilic life lends major support to the first part of the Rare Earth Hypothesis. The almost ubiquitous presence of extremophiles on

Earth in regions previously thought too hot, cold, acidic, basic, or saline shows that (at least in microbial form) life can exist in a much wider range of habitats than previously thought. This is the strongest evidence that life might be widespread in the Universe (and thus perhaps widespread in the solar system). But there is a second major implication of the discovery of extremophiles: They show that life can exist well above and below the temperature range (32–212°F) that allows for the existence of liquid water at a pressure of 1 atmosphere, the conditions found in what has been called the *habitable zone*. The extremophiles have rendered the original concept of the habitable zone obsolete. In our solar system, surface water exists only on Earth (and perhaps on Europa), so if we assume that we will find life only on planets with water, then we would have to conclude that only these two bodies should harbor life of any sort. The discovery of the extremophiles requires us to revise that thinking. Let us keep this in mind as we examine, in Chapter 2, the concept of habitable zones.

Habitable
Zones
of the Universe

The Earth would only have to move a few million kilometers
sunward—or starward—for the delicate balance of climate
to be destroyed. The Antarctic icecap would melt and flood
all low-lying land; or the oceans would freeze and the whole
world would be locked in eternal winter.

—Arthur C. Clarke, *Rendezvous with Rama*, 1973

L ocation! Location! Location! The secret for producing great Hollywood films—and for selling real estate—is also life's secret for populating the Universe. Much of the Universe is clearly hostile to life,
and only rare places offer even potential oases for its existence. Empty space,
the interiors of stars, frigid gas clouds, the "surface" of gaseous planets like
Jupiter—all must be lifeless. We cannot know for certain what the limits are
for life's environments, but looking at what is needed to support Earth life
provides a basis for estimating where in the Universe life might exist. We
speculate in this manner with the understanding that we have a biased

perspective—that of inhabitants of a planet that seems to provide a nearly perfect habitat.

One of Earth's most basic life-supporting attributes is indeed its location, its seemingly ideal distance from the sun. In any planetary system there are regions—distances from the central star—where a surface environment similar to the present state of Earth could occur. The favorable region or distance from the star is the basis for defining the "habitable zone" (referred to by astrobiologists as the HZ), the region in a planetary system where habitable Earth clones might exist. Since its introduction, the concept of habitable zone has been widely adopted and has been the subject of several major scientific conferences, including one held by Carl Sagan near the end of his brilliant career.

The defining aspect of the HZ is that it is the region where heating from the central star provides a planetary surface temperature at which a water ocean neither freezes over nor exceeds its boiling point (see Figure 2.1). The actual width of the HZ depends on how Earth-like we decide a planet must be to be deemed habitable. Extreme events, such as the loss of oceans or a deep planetary freeze, may seem totally preposterous to Earthlings happily living in nearly ideal climatic conditions, but these events would surely occur if Earth were (on the one hand) slightly closer to or (on the other) slightly farther from the sun. Occupying the HZ, or planetary "comfort zone," is analogous to sitting next a campfire on a cold night. Imagine trying to survive a night in the Yukon when the temperature is 100°F below zero. You have a large campfire, but if you sleep too close to it you catch on fire, and if you are too far back you freeze.

Astronomers held the first discussions of the habitable zone in the 1960s. The range of the habitable zone was considered to be bounded by two effects: low temperature at the outer edge and high temperature at the inner edge. Our closest neighbors in space provide sobering examples of what happens to planets close to, but not within, the HZ. Closer to the sun than the HZ, a planet gets too hot. Venus is an example. The surface of this neighbor is nearly hot enough to glow. If Venus ever had an ocean, it has long since evaporated and been totally lost to space.

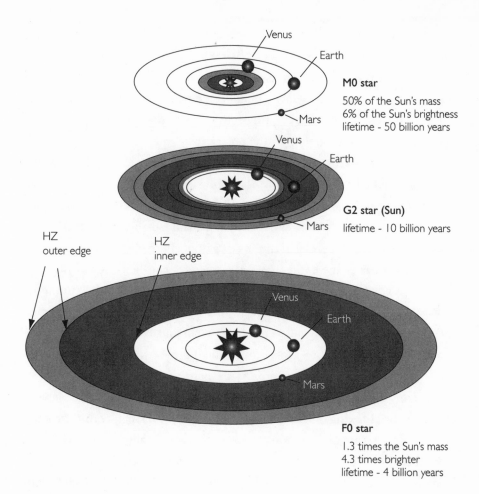

Figure 2.1   *Estimates of the habitable zones (HZ) around stars that are slightly less and slightly more massive than the Sun (based on results of Kasting, Whitmore, and Reynolds, 1993). Two estimates of the cold outer edge of the HZ are based on the temperature where CO_2 (dry ice) begins to condense in the atmosphere (inner limit) and the theory that Mars was in the Sun's HZ early in its history (the outer limit). The hot inner edge of the HZ is estimated both in terms of the belief that any oceans on Venus boiled away at least a billion years ago and in terms of estimates of the atmospheric conditions required to produce runaway greenhouse heating.*

Outside of the HZ, temperatures are too low. Mars, for example, is frozen to depths of many kilometers below its surface. If Earth were moved outward (or if the sun reduced its energy output), Earth's atmosphere would cool to a point where the planet would become ice-covered. Eventually, carbon dioxide would freeze to form reflective clouds of "dry ice" particles, and ultimately, CO_2 would freeze on the polar caps.

In 1978 the astrophysicist Michael Hart performed detailed calculations and reached a stunning conclusion. His work included the well-known fact that the sun becomes slightly brighter with time. About 4 billion years ago, the sun was about 30% fainter than at present. As the sun brightens, the HZ drifts outward. Hart called the small region wherein Earth would remain within the HZ over the entire age of the solar system the continuously habitable zone, or CHZ. His computations indicated that sometime during its history, Earth would have experienced runaway glaciation if it had formed 1% farther from the sun and would have experienced runaway greenhouse heating if it had formed 5% closer to the sun. Both of these effects were considered irreversible. Once frozen or fried, there could be no turning back. It is now considered possible that a frozen planet might become habitable with continued brightening of its central star. If the shape of Earth's orbit had been more elliptical, these limits would have been even smaller. Hart's work implied that the CHZ was astonishingly thin for the sun and that for stars of lower mass it did not even exist. This suggested that Earth-like planets with oceans and life were rare indeed.

Hart's CHZ is now believed to be too narrow because of several effects that he did not take into account. One of these is the discovery of a remarkable chemical process known as the CO_2–silicate cycle that, on Earth, acts as a regulating thermostat to keep the planetary temperature within "healthful" limits. This cycle can maintain habitable surface temperatures over a moderate range of solar heating effects. CO_2 is a trace gas that constitutes only 350 parts per million of the atmosphere, but it is a "greenhouse" gas: Its infrared-absorbing properties retard the escape of heat back into space. This greenhouse effect warms Earth's surface about 40°C above the temperature it would otherwise have. As we will see later in the book, the thermostatic control of the CO_2–silicate cycle (which is also known as the CO_2–rock cycle) occurs because of the effects of weathering. If the planet warms, increased weathering removes CO_2 from the atmosphere, and the loss of CO_2 leads to cooling. When Earth is too cool, weathering and CO_2 removal decrease, while the continual atmospheric buildup of volcanic CO_2 leads to warming. This remarkable negative-feedback system widens the continuously habitable zone and also complicates efforts to determine its boundaries precisely, because the CO_2–rock cycle is not perfectly understood on a planetary scale. Using this new information, astrobiologist James Kasting and his colleagues defined the HZ as "the region around a star in which an Earth-like planet (of comparable mass) and having an atmosphere containing nitrogen, water, and carbon dioxide is climatically suitable for surface-dwelling, water-dependent life." They estimated in 1993 that the width of the CHZ is from 0.95 to 1.15 AU (1 astronomical unit represents the distance from Earth to the sun, 93 million miles). This is much wider than Hart's estimate but still quite narrow.

The idea of a habitable zone is a very important concept of astrobiology, but being within an HZ is not an essential requirement for life. Life can exist outside the habitable zones of stars. Astronauts in an "ideally" supplied, powered, and designed spacecraft could survive almost anywhere in the solar system and (for that matter) almost anywhere in the vast, empty regions of the entire Universe. Furthermore, discovery of extremophiles requires that the HZ concept be viewed from a much different perspective than that of just a few years ago. The HZ as normally defined is really the *animal* HZ. Extremophilic organisms that live deep underground and require only minute amounts of chemical energy and water might thrive outside the HZ in a wide variety of environments, including the subsurface regions of planets, moons, and even asteroids. A good example is Europa, the moon of Jupiter that probably has a subterranean ocean. Europa may provide a fine habitat for microorganisms, even though it lies well outside the HZ as conventionally defined.

We believe that the concept of habitable zones should be expanded to include other categories. For planets like Earth, the animal habitable zone

(AHZ) is the range of distances from the central star where it is possible for an Earth-like planet to retain an ocean of liquid water *and* to maintain average global temperatures of less than 50°C. This temperature appears to be the upper limit above which animal life cannot exist (at least animal life on Earth). Because water can exist on a planetary surface at temperatures up to the boiling point, a planet with liquid water on its surface (the original criterion of the habitable zone) might be much too hot to allow animal life. The AHZ is thus a far more restricted region around a star than the HZ as used by Hart, Kasting, and other astrobiologists. An even narrower type of HZ would emerge if we wanted to consider a zone where modern humans could live— say, a planet where enough wheat or rice could be cultivated to feed several billion people. A much wider and more readily determinable HZ is the microbial habitable zone (MHZ), the region around a star where microbial life can exist. It is nearly the entire solar system, and it extends temporally from soon after formation of the planets until the present day. HZs for other major categories of life could be defined as well: The HZ for higher plants would be wider than that for animals but narrower than the HZ for microbes.

Although the habitable zone is described in terms of distance from a central star, it must also be thought of in terms of time. In the solar system, the HZs have definable widths; and as the sun constantly gets brighter, they move outward. Earth will eventually be left behind as the greenhouse effect causes it to become more like Venus. This will happen between 1 and 3 billion years from now, and Earth will have had about 5 to 8 billion years in the HZ (see Figure 2.1). For more massive stars, the evolution is much faster. For these stars the HZ is farther out and has a much shorter duration. The lifetimes of stars 50% more massive than the sun would be too short for the leisurely pace at which animal life evolved on Earth.

Biological evolution requires vast periods of time to arrive at complex organisms—periods on the order of hundreds of millions to billions of years. The AHZ and the MHZ are therefore both spatial and temporal domains. Our newly defined AHZ is obviously the most highly restrictive, but paradoxically, it also allows for the greatest diversity of life. Earth is in this zone, whereas Venus (with its hellish surface temperature) and Mars (with its frozen surface and thin atmosphere) have been outside of it for billions of years. Relative to Earth's orbit, Venus is 30% closer to the sun and Mars 50% more distant. In terms of the intensity of sunlight, the solar illumination is twice as great at Venus and only half as great at Mars.

PLANETS EJECTED OUT OF HABITABLE ZONES

As we learn more about the interactions of various stellar systems, it is becoming increasingly clear that planets are sometimes torn from the grasp of their central stars and hurtled into the darkness of space. The most common sources of such planetary ejection are interactions between giant planets. Although the orbits of the planets in our solar system have not changed appreciably for billions of years, they do interact with each other, and the shapes of their orbits do vary. Planetary systems in general are not necessarily gravitationally stable for time scales of billions of years. If Saturn were closer to Jupiter or if it were more massive, the long-term game of gravitational cat and mouse that planets play could lead to ejection of one of these planets, and it would escape into the galaxy. If Saturn were lost, then Jupiter would stay trapped in solar orbit, but its orbit would be oddly elliptical. Some of the giant planets recently discovered orbiting other stars have highly elliptical orbits, and the past ejection of a long-lost partner may have been the cause. Planets can also be ejected from binary star systems where two stars (and their planets) orbit each other.

Although it appears at first glance that ejection from a central sun would be a death sentence for any life on an ejected planet, such may not be the case. Again, the extremophilic microbes could survive in the cold of space. Such an ejected world would have no star, no orbital motion, and no "sunlight," and its surface might approach the frigid temperature of liquid helium.

Any planet ejected from a planetary system would find itself in a most bizarre situation without neighbors and without an external source of heat to warm its surface. The only thing to be seen from the surface of the planet would be the continual sweep of the stars across an eternally dark night sky. This sight would continue monotonously for billions of years. The surface of any solitary planet would cool to cryogenic temperature. Inside the planet, however, warmth would still be generated from a radioactive interior. In that case, a deep subsurface biosphere would be able to survive.

Although ejected planets might not be hospitable to life, the outlook is much more favorable for large moons orbiting ejected planets. If somehow a Jupiter with its four large moons could be ejected into interstellar space, it might provide a very interesting habitat not only for the continuation of microbial life but for its possible evolution as well. Consider life evolving on a large satellite like Europa in orbit around Jupiter. Europa is five times more distant from the sun than is Earth, so it gets only $\frac{1}{25}$ as much solar heat, which results in a surface temperature near 150 K. This is a frigid, ice-locked world that could not possibly have life on its surface. Yet in spite of its remote location, Europa is widely regarded as one of the more interesting possible environments for life in the solar system, because it probably has a warm liquid-water ocean beneath the ice. Although Europa is far from the sun, the flexing of its interior by the gravitational tidal effects of Jupiter and its other large moons generates appreciable heat. Europa has a significant ocean below a frozen ice crust, and this particular environment—if already endowed with life—could maintain itself in the cold of interstellar space.

HABITABLE ZONES
IN OTHER STELLAR SYSTEMS

The concept of habitable zones is perhaps most interesting as applied to stars other than the sun. The brightness of the star determines the location of its habitable zone, but brightness in turn depends on the star's size, type, and age.

For stars more massive than our sun, the outward migration of their HZs through time is much faster and of much shorter duration. More massive stars have shorter lifetimes. The sun should be fairly stable for nearly 10 billion years after its birth, but a star 50% more massive than the sun enters its red giant stage after only 2 billion years. When a star becomes a red giant, its brightness increases by a factor of a thousand, and the HZ retreats greatly beyond its original bounds. We have already noted that a 1.5-solar-mass star would not be around long enough for animals to evolve at the leisurely pace enjoyed by terrestrial life. More massive stars have habitable zones farther from the star—or they may have no habitable zones at all. More massive stars are hotter and radiate substantially more ultraviolet light than the sun. Ultraviolet (UV) light breaks the bonds of most biological molecules, and life must be shielded from it to survive. UV also can be disastrous for the atmospheres of Earth-like planets. It is strongly absorbed at the top of such atmospheres and is a potent high-altitude heat source than can lead to escape of the atmosphere. The sun, with its effective surface temperature of 5780 kelvin emits less than 10% of its energy in the ultraviolet range, whereas hotter stars like Sirius radiate most of their energy in the UV. Atmospheric loss may prevent terrestrial planets with oceans and atmospheres from forming around more massive stars. This atmospheric problem with planets orbiting more massive stars is in addition to limitations imposed by their shorter stellar lifetimes.

It is often said that the sun is a typical star, but this is entirely untrue. The mere fact that 95% of all stars are less massive than the sun makes our planetary system quite rare. Less massive stars are important because they are much more common than more massive ones. For stars less massive than the sun, the habitable zones are located farther inward.

The most common stars in our galaxy are classified as M stars; they have only 10% of the mass of the sun. Such stars are far less luminous than our sun, and any planets orbiting them would have to be very close to stay warm enough to allow the existence of liquid water on the surface. However, there is danger in orbiting too close to any celestial body. As planets get closer to a star (or moons to a planet), the gravitational tidal effects from the star induce synchronous rotation, wherein the planet spins on its axis only once each time it orbits the star. Thus the same side of the planet always faces the star. (Such tidal locking keeps one side of the Moon facing Earth at all times.) This synchronous rotation leads to extreme cold on the dark side of a planet and freezes out the atmosphere. It is possible that with a very thick atmosphere, and with little day/night variation, a planet might escape this fate, but unless their atmospheres are exceedingly rich in CO_2, planets close to low-mass stars are not likely to be habitable because of atmospheric freeze-out.

We can thus look at various stars in our Milky Way galaxy and ask whether they are appropriate places for life or, indeed, have habitable zones at all. For example, could there be habitable planets orbiting binary stars or multiple star systems, places where two or more stars are locked in a complex orbital dance? Can planets with stable orbits and relatively constant regimes of temperature be found in such settings? Can planets even form in such settings? These questions are highly relevant to understanding the frequency of life beyond Earth, because approximately two-thirds of solar-type stars in the solar neighborhood are members of binary or multiple star systems. Astrobiologist Alan Hale, who has written on the problems of habitability in binary or multiple star systems, notes, "The effects of nearby stellar companions on the habitability of planetary environments must be considered in estimating the number of potential life-bearing planets within the Galaxy."

Two scenarios can be considered: the case where the stellar components (the stars of the binary or multiple star system) are quite close together, and the planets orbit both or all of the stars, and the case where the stellar companions are far apart, and the planets orbit a single star. But can planets even form in such stellar systems? Some recent work suggests that planets may not be able to form there unless the stars are at least 50 times the distance from Earth to the sun or 50 AU, although this has not been proved. Alan Hale suggests that stable orbits will be achieved in multiple star systems only where the companion stars are at less than 20 million miles apart or farther than a billion miles apart. And, of course, if planets do form in such systems, two or more bodies will affect their orbits.

The most pressing question is whether planets, once formed in a multiple star system, can achieve stable orbits. The rise of life (at least on Earth) seems to require long periods of constant conditions, which require stable orbits. Highly elliptical orbits wherein a planet moves in and out of the CHZ might allow microbial life to form and even flourish but probably would be lethal to animal life. In such systems planets might form, but their orbits would be perturbed by the various gravitational forces of more than a single star, which would eventually either eject the planets or cause them to fall into one of the stars.

A second problem with multiple star systems as habitats for life is insolation (the stellar energy a planet receives). S.H. Dole, in his groundbreaking 1970 book *Habitable Planets for Man*, estimated that the average amount of energy received by a planet could vary by as much as 10% without affecting its habitability. (This too is debatable: Our sun undergoes far less variation in output than 10%, yet even these small fluctuations produce major swings in climate that drastically affect the evolution of life forms.) Where planets orbit in the same plane as the companion stars, insolation will also be affected by eclipses of one star by another.

Finally, the residents of any planet in a multiple star system will have to deal with the stellar evolution of two or more suns. Our sun is getting brighter through time. This gradual brightening causes the habitable zones to migrate ever outward. With two or more suns undergoing the same process, we might expect habitable zones to migrate even faster through time. Although this might not adversely affect microbial life, it could inhibit animal life. All in all, it appears that multiple star systems might be regions that could support life, but perhaps not animal life. They are certainly less favorable habitats for animal life than solitary stars.

Other types of stars might be even less suitable. Variable stars (those that exhibit rapidly changing insolation) are surely poor candidates for producing planets habitable by animals (though here again, microbial life might gain and maintain a foothold, assuming that planets form). Unusual stellar entities such as neutron stars and white dwarf stars are probably uninhabitable by any form of life.

What of regions where star frequency (the number of stars in a volume of space) is very high? Such regions include open star clusters and globular star clusters. Open clusters are unlikely to be hospitable to animal life because they are too young. Most are composed of relatively new stars, where life— at least advanced life such as higher plants and animals—would not yet have had a chance to develop. Many open clusters are dispersed by the time they have orbited their galaxy several times. Others are more long-lasting, but they too have problems. Because neighboring stars are so close, planetary orbits can be perturbed, causing planets to be ejected, to enter highly elliptical orbits, or even to fall into their suns.

In globular clusters the density of stars is extremely high: Some globular clusters can have as many as 100,000 stars packed into a space some tens to hundreds of light-years across. The nearest star to our own, Proxima Centauri, is 4.2 light-years away. There are a total of 23 known stars within 13 light-years of the sun. In a globular cluster, the same distance might hold 1000 stars or more. For example, the M15 globular cluster has 30,000 stars packed into a space only 28 light-years across. There would be no night on any planets in such clusters. There might be habitable stellar systems in such regions, but the very number of stars would make them more dangerous and less congenial to the maintenance of animal life than more widely separated stars; there is too much radiation and particles, too many chances for gravitational changes to affect the orbits of planets in any such mass. Being in a high concentration of stars increases the risk of a nearby star going nova (exploding) or belching hard radiation into nearby space. A second great disadvantage of globular clusters is that they are composed of old (and thus heavy-element-poor) stars, all of about the same age. The low abundance of "heavy elements" such as carbon, silicon, and iron makes it unlikely that any Earth-size terrestrial planets would form. These heavy elements are required not only to provide habitats for life but also to build life as we know it.

Even if some of the stars did manage to have Earth-like planets, the stars would be so old that 1-solar-mass stars would have evolved to the point where their HZs had retreated outward beyond the inner planets. Globular clusters thus may be devoid of all life. This conclusion illustrates real progress in our understanding of the limits of life in the cosmos. In 1974 a group of astronomers led by Frank Drake directed a radio signal toward the globular cluster M13. It was hoped that other radio-astronomers living around one of the 300,000 stars in the cluster might receive the message. Today, only a few decades later, we realize that there is no chance anyone will be there to take the call when the radio message arrives at M13, some 24,000 years from now. If the experiment were to be repeated, the beam would be directed toward stars more likely to have planets and life.

About other stellar regions we can only speculate. Stars are continuing to form: Is there some aspect of their formation that is beneficial—or deleterious—to habitability? Would a planet in a region with newly forming stars be able to sustain life? What about stellar systems in the middle of nebulae? Are these regions neutral to life, or does the presence of great quantities of interstellar gas have some effect on life's presence or existence? Our sun probably formed in a low-density star cluster that dispersed soon thereafter and thus avoided disruption of the orbits of Jupiter, Saturn, Uranus, and Neptune.

Habitable Zones in the Galaxy

The concept of habitable regions can be applied to our Milky Way galaxy as well. We (and a few other astrobiologists) suspect there are geographic regions that can be plotted from the center of our galaxy that are habitable regions in a way analogous to the habitable zones around stars. Our galaxy is a spiral galaxy (the other types are elliptical and irregular galaxies). In most galaxies the concentration of stars is highest in the center and diminishes away from the center. Spiral galaxies are dish-shaped (round, but flat if viewed from the side), with branching arms when viewed from the top. But viewed from the side they are quite flat. Our galaxy has an estimated diameter of about 85,000 light-years. Our sun is about 25,000 light-years from the center, in a region between spiral arms where star density is quite low compared to the more crowded interior. In this position we slowly orbit the central axis of the galaxy.

Like a planet revolving around a star, we maintain roughly the same distance from the galactic center, and this is fortunate. Our star—by chance—is located in the "habitable zone" of the galaxy. We suspect that the inner margins of this galactic habitable zone (GHZ) are defined by the high density of stars, the dangerous supernovae, and the energy sources found in the central region of our galaxy, whereas the outer regions of habitability are dictated by something quite different: not the flux of energy, but the type of matter to found.

At the present time, we cannot do more than crudely designate the limits of this habitable region. Its inner boundary is surely defined by celestial catastrophes occurring closer to the center, but we cannot yet estimate how close to the center of the galaxy that boundary is. Perhaps it extends 10,000 light-years from the center, perhaps more. However, we do have at least a vague idea of the forces that impose this inner limit. Life is a very complex and delicate phenomenon that is easily destroyed by too much heat or cold and by too many gamma rays, X-rays, or other types of ionizing radiation. The center of any galaxy produces all of these.

Among the lethal stellar members of any galaxy are the neutron stars called magnetars. These collapsed stars are small but astonishingly dense, and they emit X-rays, gamma rays, and other charged particles into space. Because energy dissipates as the square of distance, these objects are no threat to our planet. Closer to the center of the galaxy, however, their frequency increases. Any galactic center is a mass of stars, some the lethal neutron stars, and it seems most unlikely that any form of life as we know it could exist nearby.

An even greater threat comes from exploding stars known as supernovae. As stars grow old, they burn up their hydrogen and eventually collapse on themselves. Some of them then explode outward with terrific force. Any star going supernova would probably sterilize life within a radius of 1 light-year of the explosion and affect life on planets as far as 30 light-years away. The very number of stars in galactic centers increases the chances of a nearby supernova. Our sun and planet are protected simply by the scarcity of stars around us.

The outer region of the galactic habitable zone is defined by the elemental composition of the galaxy. In the outermost reaches of the galaxy, the concentration of heavy elements is lower because the rate of star formation—and thus of element formation—is lower. Outward from the centers of galaxies, the relative abundance of elements heavier than helium declines. The abundance of heavy elements is probably too low to form terrestrial planets as large as Earth. As we shall see in the next chapter our planet has a solid/liquid metal core that includes some radioactive material giving off heat. Both attributes seem to be necessary to the development of animal life: The metal core produces a magnetic field that protects the surface of the planet from radiation from space, and the radioactive heat from the core, mantle and crust fuels plate tectonics, which in our view is also necessary for maintaining animal life on the planet. No planet such as Earth can exist in the outer regions of the galaxy.

Not only is Earth in a rare position in its galaxy; it may also be fortunate (at least as far as having life is concerned) in being in a spiral rather than an elliptical galaxy. Elliptical galaxies are regions with little dust which apparently exhibit little new star formation. The majority of stars in elliptical galaxies are nearly as old as the universe. The abundance of heavy elements is low, and although asteroids and comets may occur, it is doubtful that there are full-size planets.

HABITABLE ZONES, AND TIMES, IN THE UNIVERSE

Because our limitations of the Universe deal with time, we must pose our question in a temporal sense: Are there *times* that are habitable in the Universe? As we will see in the next chapters, life (at least life as we know it) requires many elements that had to be created after the Big Bang (the advent of the Universe, some 15 billion years ago). Twenty-six elements (including carbon, oxygen, nitrogen, phosphorus, potassium, sodium, iron, and copper) play a major role in the building blocks of advanced life, and many others (including the heavy radioactive elements such as uranium) play an important secondary role by creating, deep within Earth, heat indirectly necessary for life. All of these elements were created within the centers of stars—often in exploding stars, or supernovae—rather than in the Big Bang itself, so they were not present in sufficient abundance for perhaps the first 2 billion years or more of the Universe. Then, the "habitable zone" of the Universe, in the sense of time, began only after its first 2 billion years. The early history of the Universe was also dominated by objects known as quasars, which would have been very dangerous.

The early Universe must have been lifeless or at least empty of advanced life, and quite remarkably, there are also limits on the time during which the Universe can exhibit Earth-like planets that provide adequate life support for advanced life. The geological activity on Earth that is so important in controlling the atmospheric temperature via the CO_2–rock cycle is driven by the heat liberated by the radioactive decay of uranium, thorium, and potassium atoms. These elements are produced by supernovae explosions, and their rate of formation is declining with time. In our galaxy, stars that form at present have less of these radioisotopes than the sun did when it formed 4.6 billion years ago. It is entirely possible that any true Earth clones now forming around other stars would not have enough radioactive heat to drive plate tectonics, a key process that helps stabilize Earth's surface temperature.

Our definition of a universe habitable zone is based on time, and though intriguing, it is still a bit unsatisfying. Is there some geographic, rather than temporal, component of the Universe that favors or is poisonous to life? If we could map the Universe, would we find favorable and unfavorable regions, just as we do in stellar systems and in our galaxy? In other words, is life uniformly distributed throughout the Universe, or are there regions where it will exist and others where it will not? We cannot yet answer questions such as this, but some remarkable new discoveries have enabled us at least to address them.

For 10 days in December 1995, the Hubble Space Telescope in orbit around Earth focused its large mirror on a small region in space. A total of 342 exposures were made in the vicinity of Ursa Major, the Big Dipper. The area of space examined is tiny: from our perspective, only $\frac{1}{30}$ the size of the full Moon. The target area in this small region—now known as the Hubble Deep Field—was a sprinkling of galaxies. The Hubble Deep Field appears to be one of the richest windows into distant galaxies known in the sky.

The results of these 10 days of photography have been nothing short of spectacular—and in a sense revolutionary. The photographs revealed galaxies 3 to 15 times fainter—and thus proportionally more distant—than any previously observed. More than 1500 individual galaxies can be identified in the photos. The light from these faint objects has come to us from the deep past—from periods long before our own galaxy formed, and our own sun. The most distant galaxies visible in these photos probably date to some time during the first few billion years after the start of the Universe, and hence they may antedate life anywhere. It is unlikely that any of the stars in these galaxies could have Earth-like planets because the heavy elements to build them were not yet abundantly available. We thus may be seeing images of the prebiotic Universe.

Another insight gleaned from the Hubble Deep Field is that older galaxies seem to have more irregular shapes than newer galaxies. From 30% to 40% of the most distant galaxies (and hence of the the oldest galaxies) are unusual or deformed compared to those nearest our own galaxy. The galaxies of the early Universe are quite different from newer galaxies. Does galaxy morphology affect habitability? And has habitability changed through time?

An even more surprising result was the finding that the various distances from Earth of the many galaxies seen in these photos cluster around a few values. Galaxies appear to be concentrated in great bubble-like or sheet-like structures with vast voids between. We might ask whether regions along these great sheets of galaxies have higher or lower hospitality to life. A key to habitability in various galaxies may be their abundance of heavy elements. Planets that form around metal-poor stars may be too small to retain oceans, atmosphere and plate tectonics. Metal-poor planets may not be able to support or maintain animal life, for reasons that we will detail in later chapters. It is known that entire galaxies are metal-poor and hence likely devoid of animal life.

THE END OF PLANETARY HABITABILITY

For nearly all of Earth history, life was limited to creatures so small that they are invisible to the naked eye. Casual inspection during all that time would have suggested that it was a failed planet. In other planetary systems, primitive life might flourish but never advance to the point where forests and flying animals even get a serious chance to evolve. Stars with short lifetimes, unstable planetary atmospheres, changes in orbital or spin axis, massive extinctions, impacts, crustal catastrophes, the cessation of plate tectonics, or any of a whole raft of other problems could prevent the evolution of advanced life or its prolonged survival. And on Earth itself, complex life has thrived only for the last 10% of the planet's existence.

Perhaps the most predictable aspect of advanced life (if it exists) on other planets around other stars is that its tenure is limited and that eventually any such life—and even some of the planets—will perish. Like individual organisms, planets and their grand environments have life spans. All planets with life eventually become extinct. This final outcome may be brought about by external sources such as impacts or a nearby supernova, by internal effects such as atmospheric or biological catastrophe, or (if all else fails) by increase in the brightness of the central star. This will be the ultimate fate of Earth: Life on our planet will eventually be roasted out of existence. The sun is slowly getting brighter. It is now 30% brighter than it was in the early history of the planet. Over the next 4 billion years it will double in brightness. Even if life survives this travail, it will soon be stilled. About 4 billion years from now, the sun will begin to expand rapidly in size, and its brightness will dramatically increase. The sun will become a red giant, as did the stars Antares in the constellation Scorpio and Betelgeuse in the constellation Orion. In a billion-year time span, its brightness will increase over 5000 times.

At the very beginning of this process, Earth's oceans will vaporize, driving our precious water supply into space. In the final stages of its transformation into a red giant, the sun will expand to the point where it will nearly reach the orbit of Earth. The Universe will be one living planet poorer.

SUMMARY

A review of habitable zones—for animals as well as microbes, and in the galaxy and Universe as well as around our sun—leads to an inescapable conclusion: Earth is a rare place indeed. Perhaps the most intriguing finding of this line of research is that Earth is rare as much for its abundant metal content as for its location relative to the sun. As we will see in the next chapter, the metal-rich core of our Earth is responsible for much of its hospitality to life.

Building
a Habitable
Earth

The Earth is the only world known, so far, to harbor life.
There is nowhere else, at least in the near future, [to which]
our species could migrate.

—Carl Sagan, *Pale Blue Dot*

Most of the Universe is too cold, too hot, too dense, too vacuous, too dark, too bright, or not composed of the right elements to support life. Only planets and moons with solid surface materials provide plausible oases for life as we know it. And even among planets with surfaces, most are highly undesirable. As we noted in the Introduction to this book, of all yet *known* celestial bodies, Earth is unique in both its physical properties and its proven ability to sustain life. The success of Earth in supporting life for billions of years is the result of a remarkable sequence of physical and biological processes; knowledge of these processes is our main source of insight into the possibilities of life elsewhere. In this chapter we will describe the formation and evolution of the planet Earth. Understanding how Earth attained its life-giving properties will provide a framework for understanding what is required for life and how likely it is to exist on other bodies.

Using Earth to generalize about what life requires is, of course, fraught with uncertainty. Lacking knowledge of any extraterrestrial life forms, we cannot be confident that we understand the optimal or even the minimal conditions necessary to support life beyond this planet. But our planet is an uncontested success in terms of the abundance and variety of life it sports, even though it was certainly sterile soon after its formation. How did that change come about, and what were the physical attributes of Earth that allowed it to become so rich with life?

Earth is the only location in the universe that is known to have life, but it is only one of perhaps millions of habitats in our galaxy, and trillions in the Universe, that might also harbor life. From the biased viewpoint of Earthlings, however, it does appear that Earth is quite a charmed planet. It has the right properties for the only type of life we know, it formed in the right place in the solar system, and it underwent a most remarkable and unusual set of evolutionary processes. Several of its neighbors in the solar system even played highly fortuitous, supporting roles in making Earth a congenial habitat for life. The near-ideal nature of Earth as a cradle of life can be seen in its prehistory, its origin, its chemical composition, and its early evolution. What are the most important factors that allowed Earth to support advanced life? Earth has offered (1) at least trace amounts of carbon and other important life-forming elements, (2) water on or near the surface, (3) an appropriate atmosphere, (4) a very long period of stability during which the mean surface temperature has allowed liquid water to exist on its surface, and (5) a rich abundance of heavy elements in its core and sprinkled throughout its crust and mantle regions.

Earth is actually the final product of an elaborate sequence of events that occurred over a time span of some 15 billion years, three times the age of Earth itself. Some of these events have predictable outcomes, whereas others are more chaotic, with the final outcome controlled by chance. The evolutionary path that led to life included element formation in the Big Bang and in stars, explosions of stars, formation of interstellar clouds, formation of the solar system, assembly of Earth, and the complex evolution of the planet's interior, surface, oceans, and atmosphere. If some god-like being could be given the opportunity to plan a sequence of events with the express goal of duplicating our "Garden of Eden," that power would face a formidable task. With the best intentions, but limited by natural laws and materials, it is unlikely that Earth could ever be truly replicated. Too many processes in its formation involved sheer luck. Earth-*like* planets could certainly be made, but each outcome would differ in critical ways. This is well illustrated by the fantastic variety of planets and satellites that formed in the solar system. They all started with similar building materials, but the final products are vastly different from each other. Just as the more familiar evolution of animal life involved many evolutionary pathways with complex and seemingly random branch points, the physical events that led to the formation and evolution of the physical Earth also required an intricate set of nearly irreproducible circumstances.

Any construction project requires that building materials be on site before the actual construction begins. The formation of Earth was no different. Hence the first step is assembling the raw materials.

CREATION OF THE ELEMENTS

Although we understandably date Earth's history from the origin of the planet, a considerable "prehistory" preceded Earth's formation. One of the most important aspects of this period was the origin of the chemical elements. The elements are the building blocks of both planets and life. Consider that in a sort of cosmic reincarnation, every atom in our bodies resided inside several different stars before the formation of our sun and has been part of perhaps millions of different organisms since Earth formed. Planets, stars, and organisms come and go, but the chemical elements, recycled from body to body, are essentially eternal.

All but a minute fraction of the atoms in the planet Earth and in its inhabitants were produced, long before Earth formed, by an intricate set of astrophysical processes. A most remarkable aspect of our prehistory is that the processes of element formation were universal and that they provided fairly similar starting materials for most planets' building processes, wherever these occurred. Planets and the life they harbor may develop great variations, but their initial stores of building blocks were similar, largely because of the relative abundance of the various chemical elements. By looking at this prehistory, we can gain insight into the range of possible planets and life habitats that might form in different places and at different times in the Universe.

The cosmic choreography that led to the formation of Earth, all other bodies in the Universe, and (ultimately) life began with the Big Bang, the very "beginning of time." The Big Bang is what nearly all physicists and astronomers believe is the actual origin of universe. Born in an instant, the entire universe started out as an environment of incredible heat and density, but subsequent expansion led to rapid cooling and more rarefied conditions. During the first half-hour, conditions existed that produced most of the atoms that are still the major building blocks of the stars—mainly hydrogen and helium, atoms that make up over 99% of the normal (visible) matter in the universe. In itself however, the Big Bang generated little chemical diversity. It gave us little or nothing beyond hydrogen, helium, and lithium to fill the periodic table. It did not produce oxygen, magnesium, silicon, iron, and sulfur, the elements that constitute more than 96% of the mass of our planet. It did not produce carbon, a chemically unique element whose versatile ability to form complex molecules is the basis for all known life. But the Big Bang did produce the raw material (hydrogen) from which all heavier and more interesting elements would later form.

The temperature of the Universe, during its first half-hour, was above 50 million degrees Celsius. At this temperature, positively charged protons (the nuclei of hydrogen) could occasionally collide with enough energy to overwhelm the electrostatically repulsive effects of their like positive charges and fuse together to form helium. This simple fusion process is the secret of the stars. It is the reason why the night sky is not dark, the reason why Earth's surface is not frozen, the reason why planets can exist; it is the energy source that powers life on Earth. This process commonly occurs inside stars, but it was also the major nuclear reaction in the Big Bang. In stars the fusion of hydrogen to yield helium provides a critical long-term energy source, but in the Big Bang, helium production was a mere footnote to the grand events that had just preceded it. In addition to being the first nuclear reaction to produce new elements, the formation of helium from hydrogen (thermonuclear fusion) has handed advanced life a double-edged sword. On the positive side, fusion is the only known process that could be used in future reactors to provide truly long-term energy sources for advanced civilizations. (Fossil fuel and solar power could not possibly supply Earth's human population, at its present rate of energy consumption, for more than a few thousand years. Fusion reactors using hydrogen from the ocean could, in principle, produce nearly unending supplies of energy.) On the other hand, bombs based on the fusion of hydrogen are one of the surest means of destroying advanced life forms on a planet-wide scale.

The fusion of hydrogen to form helium was the end of the road for element production during the Big Bang. The key process that would lead from helium to the production of heavier elements could not occur under the conditions that prevailed in the early Universe. When the temperature was high enough to produce them, the spatial density of atoms was too low and the reaction rates too small. Thus it was not possible for Earth-like planets to form in the early Universe, because their formation depends on elements heavier than helium. During the first 15% of the age of the Universe, a period of over 2 billion years, stars could form, but there was not enough dust and rocks for them to have terrestrial planets. When modern telescopes are used to observe more and more distant objects, we are actually seeing further and further back into the early history of the Universe. If it were possible to detect life with a telescope, we would observe a "dead zone" beyond a certain distance— beyond a certain time, that is, when the Universe was without life or planets or even the elements to produce them.

The trick for getting from helium to the generation of planets, and ultimately to life, was the formation of carbon, the key element for the success of life and for the production of heavy elements in stars. Carbon could not form in the early moments following the Big Bang, because the density of the expanding mass was too low for the necessary collisions to occur. Carbon formation had to await the creation of giant red stars, whose dense interiors are massive enough to allow such collisions. Because stars become red giants only in the last 10% of their lifetimes (when they have used up much of the hydrogen in their cores), there was no carbon in the Universe for hundreds of millions to several billion years after the Big Bang—and hence no life as we know it for that interval of time.

Carbon formation requires three helium atoms (nuclei) to collide at essentially the same time: a three-way collision. What actually happens is that two helium atoms collide to form the beryllium-8 isotope, and then, within a tenth of a femtosecond (1/10,000,000,000,000,000 second) before this highly radioactive isotope decays, it must collide with and react with a third helium nucleus to produce carbon. Carbon has a nucleus composed of six protons and six neutrons, the cumulative contents of three helium atoms. Once carbon had been made, however, heavier and heavier elements could be formed. The production of heavier and more interesting elements occurred in the fiery cores of stars where temperatures ranged from 10 million to over 100 million degrees Celsius. The sun is currently producing only helium, but in the future, in the last 10% of its lifetime, it will produce all of the elements from helium to bismuth, the heaviest nonradioactive element in nature. Elements heavier than bismuth are all radioactive, and most are produced by the decay of uranium and thorium. The elements heavier than bismuth were produced in the cores of stars ten times more massive than the sun that underwent supernova explosions, dramatic events in which a star brightens by a factor of 100 billion over a period of a few days.

The sequence of element production in the Big Bang and in stars provided not only the elements necessary for the formation of Earth and the other terrestrial planets but also all of the elements critical for life—those actually needed to form living organisms and their habitats. Among the most

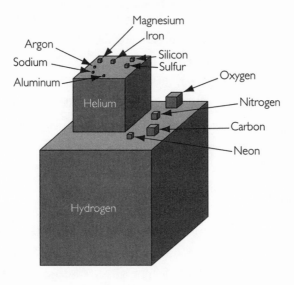

Figure 3.1   *The relative proportions (by number) of the most abundant elements in the Sun. Hydrogen and helium and the elements resting directly on top of the hydrogen cube dominate the composition of stars and Jovian planets. The terrestrial planets could not efficiently incorporate these elements and are composed largely of oxygen and the elements resting on the helium cube.*

important of these elements were: iron, magnesium, silicon, and oxygen to form the structure of Earth; uranium, thorium, and potassium to provide radioactive heat in its interior; and carbon, nitrogen, oxygen, hydrogen, and phosphorus, the major "biogenic" elements that provide the structure and complex molecular chemistry of life. The production of elements inside stars, along with continued recycling between stars and the interstellar medium, produced a relative proportion of the different elements known as the "cosmic abundance," the approximate elemental composition of the sun and most common stars. This composition is approximately 90% hydrogen and 10% helium, leavened with carbon, nitrogen, and oxygen at around 0.1% each and magnesium, iron, and silicon at roughly 0.01% each (see Figure 3.1). Earth itself exhibits similar relative abundance of iron, magnesium, and silicon and has some oxygen but can claim only trace amounts of the other cosmically abundant elements. The elements carbon, oxygen, hydrogen, and nitrogen dominate its biotic inhabitants, or life.

41

The processes that occurred during the billions of years of Earth's "pre-history" when its elements were produced are generally well understood. Elements are produced within stars; some are released back into space and are recycled into and out of generations of new stars. When the sun and its planets formed, they were just a random sampling of this generated and reprocessed material. Nevertheless, it is believed that the "cosmic abundance" mix of the chemical elements—the elemental composition of the sun—is representative of the building material of most stars and planets, with the major variation being the ratio of hydrogen to heavy elements.

The dominant atoms that formed Earth were silicon, magnesium, and iron, with enough oxygen to oxidize fully (from compounds such as MgO, magnesium oxide) most of the silicon and magnesium and part of the iron. Earth's oxygen content is 45% by weight but 85% by volume. Other elements are rare, but some play very critical roles. Carbon is a trace element in Earth, but as we have noted, it is the key element for terrestrial life, and its rich chemical properties are probably the basis of any alien life as well. Hydrogen is also a trace element in planet Earth; still its gifts include the oceans and all water, the essential fluid of terrestrial life. Other important trace elements are uranium, potassium, and thorium. The decay of these radioactive elements heats Earth's interior and fuels the internal furnace that drives volcanism, the vertical movement of matter within its interior, and the drift of continents on its surface.

The "cosmic abundance" pattern is well known in the scientific literature, but it actually isn't as "cosmic" as its name implies. It is actually the "solar abundance" pattern, because it is based on measurements of the composition of the sun and solar system. Many stars are similar in composition, but there is variation, mainly in the abundance of the heavier Earth-forming elements relative to hydrogen and helium. The sun is in fact somewhat peculiar in that it contains about 25% more heavy elements than typical nearby stars of similar mass. In extremely old stars, the abundance of heavy elements, may be as low as a thousandth of that in the sun. Abundance of heavy elements is roughly correlated with age. As time passed, the heavy-element content of the Universe as a whole increased, so newly formed stars are on the average more "enriched" in heavy elements than older ones. There are also systematic variations within the Milky Way galaxy. Stars in the center of the galaxy are richer in metals (astronomical slang for elements heavier than helium) than stars at the outer regions.

The abundance of heavy elements enters into Rare Earth considerations because it influences the mass and size of planets. If Earth had formed around a star with lower heavy-element abundance, it would have been smaller because there would have been less solid matter in the annular ring of debris from which it accumulated. Smaller size can adversely influence a planet's ability to retain an atmosphere, and it can also have long-term effects on volcanic activity, plate tectonics, and the magnetic field. If the sun were older, if it were further from the center of the galaxy, or even if it were a typical one-solar-mass star (equal to the mass of the sun), then Earth would probably be smaller. If Earth were just a little smaller, would it have been able to support life for long periods of time?

Of all these properties of the solar system, perhaps the most curious—and at the same time the least appreciated—is that it is so rich in metals. Recent studies by Guillermo Gonzalez and others have shown that the sun is quite rare in this respect. Metals are necessary attributes of planets: Without them there would be neither magnetic fields nor internal heat sources. And metals may also be a key to the development of animal life: They are necessary to important organic constituents of animals (such as copper and iron blood pigments). How did we get our surplus treasure trove of metals?

CONSTRUCTION OF PLANET EARTH

The matter produced in the Big Bang was enriched in heavier elements by cycling in and out of stars. Like biological entities, stars form, evolve, and die. In the process of their death, stars ultimately become compact objects such as white dwarfs, neutron stars, or even black holes. On their evolutionary paths to these ends, they eject matter back into space, where it is recycled and further enriched in heavy elements. New stars rise from the ashes of the old. This is why we say that *each* of the individual atoms in Earth and in all of its creatures—including us—has occupied the interior of *at least* a few different stars. Just before the sun formed, the atoms that would form Earth and the other planets existed in the form of interstellar dust and gas. Concentration of this interstellar matter formed a nebular cloud, which itself then condensed into the sun, its planets, and their moons.

Let's take a closer look at what happened. The formation process began when a mass of interstellar material became dense and cool enough to grow unstable and gravitationally collapse into itself to form a flattened, rotating cloud—the solar nebula. As the nebula evolved, it quickly assumed the form of a disk-shaped distribution of gas, dust, and rocks orbiting the proto-sun, a short-lived juvenile state of the sun when it was larger, cooler, and less massive and was still gathering mass. The planets formed from this nebula, even though the nebula itself existed for only about 10 million years before the majority of its dust and gas either formed large bodies or was ejected from the solar system.

It would be highly informative to examine similar nebulae around other young stars, but their distance from us is so great, and their size so small, that their details cannot yet be directly imaged with telescopes. Ground-based and space-borne telescopes have, however, revealed several lines of evidence suggesting that disks surround newly forming stars. Among this evidence is a peculiar and spectacular phenomenon that has only recently begun to be understood. Young stars show jets of material radiating away from them. These "bipolar nebulae" are gaseous objects resembling two giant turnips, each with its apex pointing toward the star. The jets appear to be gas ejected perpendicular to disks that apparently exist around the central star. Thus as stars form, they paradoxically also eject matter back into space. The presence of a disk in the equatorial plane of the star forces the ejected material into jets along the polar axes of the spinning system of star and disk.

In the solar nebula, 99% of the mass was gas (mostly hydrogen and helium), and the heavier elements that could exist as solids made up the remaining 1%. Some of the solids were surviving interstellar dust grains; others were formed in the nebula by condensation. This gas played a major role in forming the sun, Jupiter, and Saturn. All of the other planets, the asteroids, and the comets formed primarily from the solids. Solids were only a trace component of the nebula as a whole, but they could undergo a concentration process that gas could not. As the nebula evolved, dust, rocks, and larger solid bodies separated from the gas and became highly concentrated, forming a disk-like sheet in the mid-plane of the solar nebula, in some ways resembling the rings of Saturn.

One of the fundamental processes that led to the production of planets was accretion, the collision of solids and their sticking to one another to form larger and larger bodies. This complex process involved the formation, evolution, destruction, and growth of vast numbers of bodies ranging in size from sand grains to planets. Most of the mass of a planet was accreted from materials in its "feeding zone"—a ring section of the solar nebula disk that extended roughly halfway to the nearest neighboring planets. If viewed from above, the concentric feeding zones could be imagined as a target, with one planet forming in each radial band. The composition of solids varied with distance from the sun, so the nature of each planet was critically influenced by its feeding zone.

The accretion process was responsible for unique and very important aspects of Earth. An enigma of Earth's formation is its composition and particular location in the solar system. As we saw in Chapter 2, Earth formed within the habitable zone of the sun. A grand paradox of terrestrial planets is that if they form close enough to the star to be in its habitable zone, they typically end up with very little water and a dearth of primary life-forming elements such as nitrogen and carbon, compared to bodies that formed in the outer solar system. In other words, the planets that are in the right place, and thus have warm surfaces, contain only minor amounts of the ingredients necessary for life. The accretion process accumulated solids from the nebula, but the composition of solid dust, rocks, and planetesimals in the nebula varied with distance from the sun. At Earth's distance from the center of the solar nebula (see Figure 3.2), the temperature was too high for abundant carbon, nitrogen, or water to be bound in solid materials that could accrete to form

45

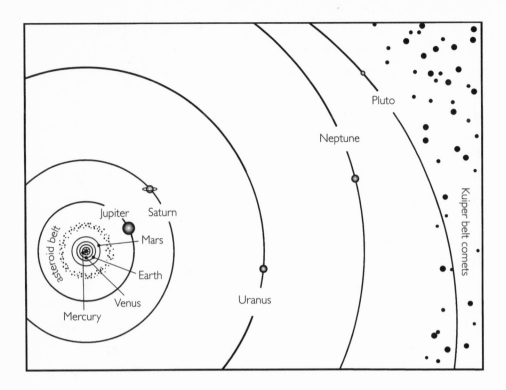

Figure 3.2   *The plan of the solar system. The rhythmic geometry of planetary spacings results from planet formation from annular "feeding zones." The asteroid belt and the Kuiper belt of comets are regions where planet growth processes failed and original planetesimals are still preserved. The planet orbits are shown to scale, illustrating how Earth and the other terrestrial planets occupy only the tiny central portion of the solar system. (The sizes of the planets are magnified by a factor of 1000; otherwise, they could not be seen at the scale of planetary orbits.)*

planetesimals and planets. Ice and carbon/nitrogen-rich solids were too volatile and had no means of efficiently forming solids in the warm inner regions of the nebula. Thus Earth has only trace amounts of these volatile components, compared to bodies that formed farther from the sun. An excellent example is the case of the carbonaceous meteorites, thought to be samples of typical asteroids formed between Mars and Jupiter. These bodies contain up to 20% water (in hydrous minerals similar to talc) and up to 4% carbon. The bulk of Earth, by comparison, is only 0.1% water and 0.05% carbon.

Had Earth formed from materials similar those in the asteroid belt, farther from the Sun, its ocean could have been hundreds of kilometers deep, and its carbon content would have been higher by many orders of magnitude. Both of these aspects would have resulted in a planet totally covered by water and with vast amounts of CO_2 in its atmosphere. The resulting greenhouse heating would have produced Venus-like surface temperatures of hundreds of degrees Celsius, and the surface would have been too hot for the complex organic molecules used by living organisms to survive. Such a planet could have developed more Earth-like conditions only if cataclysmic changes had resulted in the loss to space of most of its oceans and most of its carbon dioxide, and this seems highly unlikely. With even twice as much water, Earth would have ended up as an abyssal planet entirely covered with deep blue water—a true "water world"—and very few nutrients would have been available in the energy-rich surface waters of the ocean.

If natural processes in the nebula had acted in a different way, a radically different Earth might have resulted. For example, the reason why Earth is so carbon-poor is that most of the carbon in the inner parts of the nebula was in the form of carbon monoxide gas. Like hydrogen and helium, gaseous components could not be incorporated. If a way had existed to convert gaseous carbon into solids, then enormous amounts of carbon could have been accreted, and carbon would have been the dominant Earth element. In the cosmic abundance distribution, carbon is half as abundant as oxygen and ten times as abundant as iron, magnesium, and silicon. A genuinely carbon-rich planet would be entirely different from Earth. Imagine a planet with graphite on its surface and diamond and silicon carbide in its interior. Neither of these compounds would allow volcanism or even chemical weathering. Carbon-rich planets are presumably rare, but they probably do occur in exotic planetary systems where oxygen was less abundant than carbon in the planet-forming nebula.

The arrival of the "biogenic elements" on Earth is a matter of considerable speculation, but it is likely that most of them came from the outer regions. In the coldest outer regions of the nebula, water and nitrogen and carbon compounds could condense to form solids. Presolar interstellar solids carrying the light elements were also preserved in this region. Although most of these materials stayed in the outer solar system, some would ultimately have reached Earth by scattering. When they passed near an outer planet, their orbits about the sun could have been significantly altered, sometimes sending them toward the sun, where they might collide with terrestrial planets. Such gravitational effects from encounters with planets can cause asteroidal and cometary debris, rich in light elements, to assume earth-impacting orbits. This "cross-talk" caused some degree of mixing between different feeding zones and provided a means of bringing the building blocks of life to what might otherwise have been a lifeless planet lacking in many biogenic elements because it formed too close to the sun.

The formation of the giant outer planets is thought to have been particularly effective in scattering volatile-rich planetesimals from the outer regions of the solar system into the inner solar system, the realm of the terrestrial planets. Even today, material from the outer solar system impacts Earth. Most of the mass is in particles a quarter-millimeter in diameter that are derived from comets and asteroids. These materials carry not only carbon, nitrogen, and water but also relatively large amounts of organic material, as was first proved when extraterrestrial amino acids were discovered in the Murchison meteorite that fell in Australia in 1969. Life on Earth formed from organic compounds, and it is possible that prebiotic compounds from the outer solar system stimulated the first steps toward the origin of life on Earth. Thus the outer solar system not only provided the essential elements for life but also may have given the complex organization of the chemical processes of life a critical head start. (In the context of Rare Earth, this "seeding" would not have been unusual for a terrestrial planet. It is reasonable to expect that inner planets in all planetary systems are exposed to organic-rich "manna" from the distant comet cloud systems that invariably surround their central star.)

The scattering process that bestowed on Earth life-giving material from the outer solar system also has a dark side. We have noted that the accretion process never really ended. The rate is many orders of magnitude less than it was 4.5 billion years ago, but, as in any solar system where planets form by accretion of solids, the process still goes on. The annual influx of outer solar system material falling to Earth is 40,000 tons per year. This is mostly in the form of small particles, but larger objects occasionally hit. The small-particle flux is one 10-micron particle per square meter per day and one 100-micron particle per square meter per year. The diameter of a typical human hair is just under 100 microns. Larger objects are increasingly rare, but on the average, an outer solar system object 1 kilometer in diameter randomly impacts Earth every 300,000 years. Collision with a body this size, traveling at a speed of well over 10 kilometers per second results in a very energetic impact event. Every 100 million years, on average, a 10-kilometer object strikes Earth. Such an impact can produce a transient crater tens of kilometers deep and over 200 kilometers in diameter. It can eject enough fine debris into the air to block sunlight from the entire Earth for months. Just such an impact event killed all the dinosaurs on Earth 65 million years ago.

Early in the history of the solar system, the impact rate of very large objects was much higher, and objects struck Earth that were as big as Mars (about half the diameter of Earth). During the first 600 million years of Earth's history, there were impacts of bodies 100 kilometers in size that individually delivered enough energy to heat and sterilize Earth's surface down to depths of several kilometers. The larger impacts would have vaporized the ocean and parts of the crust. The occurrence of impacts that could cause global sterilization raises an intriguing possibility: There may have been occasions when all life on Earth was destroyed by a single impact. The intervals between devastating impacts might have been long enough for life to form and again be annihilated. If life forms easily and quickly when the conditions are right, then life might have formed and been destroyed several times before the era of sterilizing 100-kilometer and larger bodies finally ended. This effect has been called "the impact frustration of the origin of life," because life could not permanently exist on Earth until major impacts had ceased. The giant impacts essentially ended 3.9 billion years ago because most of the large impactors had been swept up by planets, ejected from the solar system, or stored in distant orbits. Over the past 3.9 billion years, impacts have continued, but not by bodies as large as 100 kilometers. Current impactors are comets and asteroids perturbed, by gravitational effects of planets, from their reservoirs in the asteroid and comet belts. The largest of these bodies can have calamitous effects (the impact of a 10-kilometer body probably caused the extinction of the dinosaurs), but they are too small to pose a threat of sterilizing the entire Earth.

The final stages of Earth's assembly process included the impact of several very large objects. In Earth's feeding zone, many celestial bodies were all struggling to grow. During accretion, a given body in a feeding zone would meet one of the following fates:

Growth by assimilation of others
Destruction by high-speed collision
Assimilation into a larger body
Ejection out of the feeding zone

The process resembled a brutal biological competition, and in the end, only a single body survived to become Earth. In the final assembly stages, however, many large bodies orbited within the feeding zone, some as large as the planet Mars. The dramatic collision of these large bodies with the young Earth played a role in determining the initial tilt values of Earth's spin axis, the length of the planet's day, the direction of its spin, and the thermal state of its interior. It is widely believed that the impact of a Mars-sized body was responsible for formation of the Moon, an oddly large satellite relative to the size of its mother planet.

The final composition of Earth had several crucial structural effects. First, enough metal was present in the early Earth to allow formation of an iron- and nickel-rich innermost region, or core, that is partially liquid. This enables Earth to maintain a magnetic field, a valuable property for a planet sustaining life. Second, there were enough radioactive metals such as uranium to make for a long period of radioactive heating of the inner regions of the planet. This endowed Earth with a long-lived inner furnace, which has made possible a long history of mountain building and plate tectonics—also necessary, we believe, to maintaining a suitable habitat for animals. Finally, the early Earth was compositionally able to produce a very thin outer crust of low-density material, a property that allows plate tectonics to operate. The thicknesses and stability of Earth's core, mantle, and crust, could have come about only through the fortuitous assemblage of the correct elemental building blocks.

There is no direct information about Earth's early history because no rocks older than 3.9 billion years have survived. We can say with confidence, however, that the period included episodes of great violence due to the effect of giant impacts. The largest high-velocity collisions would cause heating and actually resurface the planet. Cratering events of the magnitude of those that created the major basins on the Moon (the large circular regions seen without a telescope, including the eye of the "Man in the Moon") may have blown parts of the atmosphere into space. These events may have produced truly horrific environments. Impacts that vaporize large amounts of water and liberate carbon dioxide from surface rocks can lead to phenomenal greenhouse effects. After the direct heating effects due to the kinetic energy of impact have dissipated, the greenhouse gasses linger in the atmosphere and retard the escape of infrared radiation. With its main cooling process blocked, the atmosphere heats up. The greenhouse effect in the dense carbon dioxide atmosphere on modern Venus produces surface temperatures of 450°C; it has been estimated that the vast amount of gas injected into Earth's early atmosphere by giant impacts may have produced surface temperatures hot enough to melt surface rocks!

It is the heritage of terrestrial life that violent events and truly hostile environments preceded it. The violent events of these times may have determined the final abundance of water and carbon dioxide, two compounds that play crucial roles in the ability of Earth to maintain an environment where life can survive. It is interesting to speculate about what would have happened if the final abundance of these had varied. If Earth had had just a little more water, continents would not extend above sea level. Had there been more CO_2, Earth would probably have remained too hot to host life, much like Venus.

FINISH WORK

A process that strongly influenced the ultimate evolution of life on Earth was the creation of the atmosphere, oceans, and land. Events involved in the formation of all three were highly intertwined.

Without an atmosphere there would be no life on Earth. Its composition over Earth history is one of the reasons why our planet has remained a life-supporting habitat for so long. Today the atmosphere is highly controlled by biological processes, and it differs greatly from those of other terrestrial planets, which range from essentially no atmosphere (Mercury) to a CO_2 atmosphere a hundred times denser (Venus) and a CO_2 atmosphere a hundred times less dense (Mars). Even viewed from a great distance, Earth's strange atmospheric composition would provide a strong clue that life is present. Composed of nitrogen, oxygen, water vapor, and carbon dioxide (in descending order of abundance), it is not an atmosphere that could be maintained by chemistry alone. Without life, free oxygen would rapidly diminish in the atmosphere. Some of the O_2 molecules would oxidize surface materials, and others would react with nitrogen, ultimately forming nitric acid. Without life, the CO_2 abundance would probably rise, resulting in a nitrogen and CO_2 atmosphere. To an alien astronomer, Earth's atmospheric composition would be clearly out of "chemical equilibrium." This situation would provide convincing evidence of life and a vigorous ecosystem capable of controlling the chemical composition of the atmosphere. Telescopic detection of such peculiar atmospheres is the basis of a strategy for detection of life outside the solar system by the "Terrestrial Planet Finder"; we will discuss it in Chapter 10.

The atmosphere was formed by outgassing from the interior, a process that released volatiles originally carried to Earth in planetesimal bodies as well as by delivery from impacting comets. The composition and density of the atmosphere are influenced by the amount and nature of the original accreted materials, but in Earth's case they are most strongly affected by processes that recycle atmospheric components in and out of the atmosphere.

The oceans are a by-product of outgassing and the formation of the atmosphere. When the atmosphere was very hot, a great deal of it was composed of steam. Gradually, as the early Earth cooled, the steam condensed as water and formed the vast oceans we still see today. Although they were originally fresh, the oceans became salty through chemical interactions with Earth's crust.

Land provides a home for nonaquatic life, and the vast regions of shallow water that surround land offer crucial and complex habitats where oceanic life can flourish. Shallow water is also a setting where interactions between ocean and atmosphere alter the composition of the atmosphere. Earth's topography and the total amount of water determine what fraction of Earth's surface is land. The oceans contain enough water to cover a spherical Earth to a depth of about 4000 meters. If the surface of the planet varied only a few kilometers in elevation, Earth would be devoid of land. It is easy to imagine an Earth covered by water, but it is difficult to imagine that, with its present water supply, it could ever be dominated by land. To make more land or even produce an Earth dominated by land, the oceans would have to be deeper to accommodate the same volume of water in spite of having less total surface area. Thus the planet's remarkable mixture of land and oceans is a balancing act.

Land formation on Earth has throughout its history occurred by two principal means: simple volcanism creating mountains and the more complex processes related to plate tectonics. Simple volcanism leads to the formation of small islands such as Hawaii and the Galapagos archipelagos. Volcanic islands similar to Hawaii were probably the predominant landform on the early Earth. These were lifeless islands with no plant roots to slow the ravages of erosion. Low islands would have been bleak and desert-like, sterile surfaces bombarded by intense ultraviolet radiation from the sun unfiltered by Earth's early atmosphere. If climatic conditions were anything like today, the higher islands would have had copious rainfall leading to extensive erosion. Although Earth evolved beyond the stage where its only land consisted of eroding and doomed islands, many of the water-covered planets elsewhere probably only have transient basaltic islands, at best. At worst they have no land at all.

In the case of our planet, Earth managed to form continents that could endure for billions of years. This required the formation of land masses made of relatively low-density materials that could permanently "float" on the denser underlying mantle while parts of them extended above the sea.

How did the first continents form? Early continental land masses may have formed when the impact of large comets and asteroids melted the outer region of Earth to form a "magma ocean," a planet-smothering layer of molten rock. The concept of a global magma ocean grew out of studies of the Moon. The heat generated from the rapid accretion of many planetesimals into our solid Earth appears have melted the upper 400 kilometers of the Moon's surface. In the lunar case, as the magma ocean cooled, myriad small crystals of a mineral called plagioclase feldspar (a low-density mineral rich in calcium, aluminum, and silicon) formed and floated upward to create a low-density crust nearly 100 kilometers thick. This ancient crust is still preserved and can even be seen with the naked eye as the bright, mountainous lunar "highlands." In like fashion, a magma ocean on Earth, may have led to formation of the first continents. Alternatively, the processes leading to the formation of the first continental land may have occurred beneath large volcanic structures. The initial land mass was small and was not until half way through its history that land covered more than 10% of the Earth's surface. In any event, the outcome was a planet with both land and sea. This fortuitous combination may be the most important factor that ultimately made life possible.

By about 4.5 billion years ago Earth was built. The next step was to populate it—the subject of the next chapter.

Life's
First Appearance
on Earth

One amino acid does not a protein make—let alone a being.

—Preston Cloud, *Oasis in Space*

Once life evolves, it tends to cover its tracks.

—John Delaney

The discovery of extremophilic microbes has radically changed our conception of where life might be able to exist in the Universe—it causes us to reassess the concept of habitable zones. Scientists now realize that habitats suitable for *microbial* life are far more widely distributed in our solar system, and surely in the Universe as well, than was considered possible even in the most optimistic views of the 1980s and before. On the other hand, these same studies are showing that complex life—such as higher animals and plants—may have fewer suitable habitats than was previously thought. But just because life *could* exist in a place doesn't mean it is actually there. Life can be widely distributed in the Universe only if it can come into

55

being easily. In this chapter we will examine current knowledge and hypotheses about how life may have first formed on Earth and in what type of environment this may have taken place.

How Did Life Begin?

What really is life? And how do we recognize its formation? These questions seem simple, but the answers are dauntingly complex. In its most common-sense definition, life is able to grow, reproduce, and respond to changes in the environment. By this definition, extremophiles are obviously alive, for example. Yet many crystals can do as much, and they are clearly not life. The great British biologist J.B.S. Haldane pointed out that there are about as many living cells in a human being as there are atoms in a cell. Individual atoms themselves are not alive, though. "The line between living and dead matter is therefore somewhere between a cell and an atom," Haldane concluded.

Somewhere in between atoms and the living cell there is the entity known as a virus. Viruses are smaller than the smallest living cells and do not seem to be alive when isolated (they cannot reproduce), yet they are capable of infecting and then changing the internal chemistry of the cells they invade. Are they alive? In isolation they do not seem to be, but in combination with the host they very well may be. The boundary between living and nonliving is ambiguous at these levels of organization. By the time we reach the level of organization found in bacteria and archaea, however, we are sure that we have unambiguous life. We are also sure that all life on Earth is based on the DNA molecule.

Deoxyribonucleic acid, or DNA, is predominantly composed of two backbones that spiral around one another (the famous "double helix" described by its discoverers, James Watson and Francis Crick). These two spirals are bound together by a series of projections, much like steps on a ladder, made up of the distinctive DNA bases adenine, cytosine, guanine, and thymine. The term *base pair* comes from the fact that the bases always join up in the same way: Cytosine always pairs with guanine, and thymine always joins with adenine. The order of bases on each strand of DNA supplies the language of life; these are the genes that code for all information about a particular life form.

There might be many kinds of life elsewhere in the Universe, and there is a great deal of speculation among scientists about whether DNA is the only molecule on which life can be based or one of many. It is certainly the only one capable of replication and evolution on Earth, and all life here contains DNA. The fact is that all organisms on Earth share the same genetic code is the strongest evidence that all life here derives from one common ancestor.

Was the rise of life inevitable on this planet? Let us perform a thought experiment: If every environmental condition that ever existed on the Earth during its 4.5-billion-year history were exactly reduplicated in the same order, would life itself again evolve? And if it did, would it evolve with DNA as its crux?

The formation of this complex molecule is thus the starting point for any discussion of life's history on this planet—and perhaps on any other. There may indeed be other ways to produce life; one would be a system where ammonia, rather than water, is the solvent necessary for life. This route may even have been followed, only to be erased later, probably because water is a better solvent than ammonia. (Solvents are a rather humdrum yet essential ingredient in life's recipe. Many of the chemicals necessary for life can be delivered into a cell only in solution, and for that a solvent is needed.) Thus "DNA life" may be either the only type of life that formed or the only survivor.

Life seems to have appeared on this planet somewhere between 4.1 and 3.9 billion years ago, or some 0.5 to 0.7 billion years after Earth originated. However, the fact that no fossils were preserved at this time in Earth's history clouds our understanding of life's earliest incarnation. The oldest fossils that we do find are from rocks about 3.6 million years of age, and they look identical to bacteria still on Earth today. There may have been earlier types of life that are no longer represented on Earth, but our present knowledge suggests that bacteria-like forms were the first to fossilize.

The Earth formed about 4.5 to 4.6 billion years ago from the accretion of variously sized "planetesimals," or small bodies of rock and frozen gases. For the first several hundred million years of its existence, a heavy bombardment of meteors pelted the planet with lashing violence. Both the lava-like temperatures of Earth's forming surface and the energy released by the barrage of incoming meteors during this heavy bombardment phase would surely have created conditions inhospitable to life. As we recounted in the last chapter, this constant rain of gigantic comets and asteroids would have driven temperatures high enough to melt surface rock. No water would have formed as a liquid on the surface. Clearly, there would have been no chance for life to form or survive on the planet's surface. It was hell on Earth.

As we showed earlier, the new planet began to change rapidly soon after its initial coalescence. About 4.5 billion years ago, Earth began to differentiate into different layers. The innermost region, a core composed largely of iron and nickel, became surrounded by a lower-density region called the mantle. A thin crust of still lesser density rapidly hardened over the mantle, while a thick, roiling atmosphere of steam and carbon dioxide filled the skies. In spite of its being waterless on the surface, great volumes of water would have been locked up in Earth's interior, and water would have been present in the atmosphere as steam. As lighter elements bubbled upward and heavier ones sank, water and other volatile compounds were expelled from the interior and added to the atmosphere.

The heavy bombardment by comets and asteroids lasted more than half a billion years and finally began to diminish around 3.8 billion years ago as the majority of debris was incorporated into the planets and moons of our solar system. During the period of heaviest impact, the steady bombardment would have scarred our planet by craters in the same manner as the moon. Yet the comets and asteroids raining in from space delivered an important cargo with each blow. Some astronomers believe that much, or even most, of the water now on our planet's surface arrived with the incoming comets; others think that only a minority of Earth's water arrived in this fashion.

Comets are made up of dust and volatiles, such as water and frozen carbon monoxide, and there is no doubt that a good many of them hit the early Earth. These cargoes of water slamming into Earth would have turned instantly to steam. The dense early atmosphere remained hot for hundreds of millions of years. Perhaps 4.4 billion years ago, its surface temperatures might have dropped sufficiently, and for the first time liquid water would have condensed from steam onto our planet's surface, successively forming ponds, lakes, seas, and finally a planet-girdling ocean. The study of ancient sedimentation suggests that by slightly less than 3.9 billion years ago, the amount of oceanic water on Earth may have approached or attained its present-day value. But these were not tranquil oceans or oceans even remotely similar to those of today.

We have only to look at the Moon to be reminded of how peppered Earth and its oceans were during the period of heavy impact, between 4.4 and 3.9 billion years ago. Each successive, large-impact event (caused by comets larger than 100 kilometers in diameter) would have partially or even completely vaporized the oceans. Imagine the scene if viewed from outer space: the fall of the large comet or asteroid, the flash of energy, and the evaporation of Earth's planet-covering ocean, to be replaced by a planet-smothering cloud of steam and vaporized rock heated (at least for some decades or centuries) far above the boiling point of liquid water. It is difficult to conceive of life—whatever its form—surviving anywhere on the planet during such times, unless that survival occurred deep underground.

Scientists have made mathematical models of such ocean-evaporating impact events. The collision with Earth of a body 500 kilometers in diameter results in an almost unimaginable cataclysm. Huge regions of Earth's rocky surface are vaporized, creating a cloud of "rock-gas" several thousand degrees in temperature. It is this superheated vapor, in the atmosphere, that causes the entire ocean to evaporate into steam. Cooling by radiation into space would take place, but a new ocean produced by condensing rain would not fully form for thousands of years after the event. Much of the revolutionary detective work behind these conclusions was described in 1989 by Stanford University scientist Norman Sleep, who realized that the impact of such a large asteroid or comet could evaporate an ocean 10,000 feet deep, sterilizing Earth's surface in the process.

How ironic that the comets may have brought some of Earth's life-giving, liquid water—a prerequisite for life—and then snatched that gift away for a time with each successive large-impact event. Yet it is not only water that these comets may have brought. They could have played a role in determining the chemical evolution of Earth's crust. And they may have contributed another ingredient to the recipe for what we call life: They may have brought organic molecules—or even life itself—onto our planet's surface for the first time.

If some time machine made it possible to visit the Earth of about 3.8 billion years ago, at the end of the period of heavy bombardment, our world would surely still appear alien to us. Even though the worst barrage of meteor impacts would have passed, there still would have been a much higher frequency of these violent collisions than in more recent times. The length of the day was shorter, because Earth was rotating far faster than it does now. The sun was much dimmer, perhaps a red orb supplying little heat, for it not only was burning with less energy than today but also had to penetrate a poisonous, turbulent atmosphere composed of carbon dioxide, hydrogen sulfide, steam, and methane. In such an environment, we would have had to wear spacesuits of some sort, for only traces of oxygen were present. The sky itself might have been orange to brick red in color, and the seas, which surely covered all of the planet's surface except for a few scattered, low islands, would have been muddy brown and clogged with sediment. Yet perhaps the greatest surprise to us would be the utter absence of life. No trees, no shrubs, no seaweed or floating plankton in the sea; it would have seemed a sterile world. Somehow, the fact that we have not yet detected life on Mars seems consistent with its satellite images. A waterless world fits our picture of a lifeless world. Even when the young Earth was covered with water, however, it was still devoid of life. But not for long.

A RECIPE FOR LIFE

Most scientists are confident that life had already arisen 3.8 to 3.9 billion years ago, at about the time when the heavy bombardment was coming to an end. The evidence indicative of life's appearance is not the presence of *fossils* but the isotopic signatures of life extracted from rocks of that age in Greenland.

The oldest rocks on Earth that have been successfully dated via radiometric dating techniques are mineral grains of zircon about 4.2 billion years old. The Greenland rocks (from a locality called Isua) are thus only slightly younger. The Isua rock assemblages, which include sedimentary (or layered) rocks as well as volcanic rocks, have yielded a most striking discovery. They contain ratios of the light and heavy isotopes of carbon, indicating formation in the presence of life. The isotopic residue in the Isua rocks reveals an excess of the isotope carbon-12 compared to carbon-13. A surplus of carbon-12 is found today in the presence of photosynthesizing plants, because all living organisms show an enzymatic preference for "light" carbon. The inference is that if early life existed at Isua, it may have used photosynthesis for its energy sources. But there is no fossil evidence that life existed so long ago—only this enigmatic and provocative surplus of a carbon isotope that in our day is a sign of life's presence. If the excess of light-carbon isotope is indeed a reliable indication that ancient life existed at Isua, and perhaps elsewhere on Earth, as early as 3.8 billion years ago, it leads to a striking conclusion: Life seems to have appeared simultaneously with the cessation of the heavy bombardment. As soon as the rain of asteroids ceased and surface temperatures on Earth permanently fell below the boiling point of water, life seems to have appeared. But how?

There are still more questions than answers about life's origin on Earth. Yet the sophistication of the questions now being addressed by legions of scientists tells us that we are well along in the investigation. Among the most pressing of these questions: Did life originate in only a single setting or in several? Did the key chemical components—the building blocks—come from different environments to be assembled in one place? Was life's origin "deterministic"? That is, could different environmental conditions produce the same molecule of life, the familiar DNA? Were the individual stages in the origin of life (such as the formation of amino acids, then of nucleic acids, and then of cells) dependent on long-term changes in the Earth environment? Did the origin of life change the environment such that life could never originate again? At what stage did evolution take over to influence the development of life? And, perhaps most interesting of all, can we infer the nature of the settings of life's origin from the study of extant organisms—creatures living on Earth today?

Determining how the first DNA molecules appeared on Earth has been a very difficult scientific problem, and it is still far from solved. No one has yet discovered how to combine various chemicals in a test tube and arrive at a DNA molecule. Added to this is the fact that conditions on the early Earth would have been in many ways horrific for natural "chemistry" experiments involving reactions that now routinely take place in what we humans call "room temperature." Temperatures on the early Earth even 3.8 billion years ago, about the time that the first life on Earth may have appeared, may have been far higher than those of today (although some astrobiologists think that Earth was *colder* then than it is now because the sun was fainter). Many other aspects of early environmental conditions would clearly have been deleterious to much of the life now on our planet. For example, with an oxygen-free atmosphere the amount of ultraviolet radiation reaching Earth's surface would have been far higher than it is today, making delicate chemical reactions on the planet's surface very difficult. But we know that life did arise and that the most important step in the process was the formation of DNA, life's basic information center.

To build DNA—and, ultimately, life—requires the following ingredients and conditions: energy, amino acids, factors that make chemical concentration possible, catalysts, and protection from strong radiation or excess heat. The chemical evolution of life entails four steps:

1. The synthesis and accumulation of small organic molecules, such as amino acids and molecules called nucleotides. The accumulation of chemicals called phosphates (one of the common ingredients in plant fertilizer) would have been an important requirement, because these are the "backbone" of DNA and RNA.

2. The joining of these small molecules into larger molecules such as proteins and nucleic acids.

3. The aggregation of the proteins and nucleic acids into droplets that took on chemical characteristics different from their surrounding environment.

4. The replicating of the larger complex molecules and the establishment of heredity. The DNA molecule can accomplish both, but it needs help from other molecules, such as RNA.

RNA molecules are similar to DNA in having a helix and bases. But they differ in having but a single strand, or helix, rather than the double helix of DNA. They also differ in the makeup of their base composition: Instead of thymine, they contain a base called uracil. Most RNA is used as a messenger, sent from DNA to the site of protein formation within a cell, where the specific RNA provides the information necessary to synthesize a particular protein. To do this, a DNA strand partially unwinds, and an RNA strand forms and keys into the base-pair sequence on the now-exposed DNA molecule. This new RNA strand matches with the base pairs of the DNA and, in so doing, encodes information about the protein to be built. This process is called translation.

BUILDING CODE

Some of the steps leading to the synthesis of DNA and RNA can be duplicated in the laboratory; others cannot. We have no problem creating amino acids, life's most basic building block. Researchers can even produce proteins, or chains of amino acids, under laboratory conditions, as was first demonstrated by University of Chicago chemists Stanley Miller and Harold Urey in a famous experiment conducted in 1952. In a scene reminiscent of some Frankenstein movie, they created the building blocks of life for the first time—in a test tube. But it has turned out that the challenge of making amino acids in the lab is trivial compared to the far more difficult proposition of creating DNA artificially. The problem is that complex molecules such as DNA (and RNA) cannot simply be assembled in a glass jar by combining various chemicals. Such organic molecules also tend to break down when heated, which suggests that their first formation must have taken place in an environment with moderate, rather than hot, temperatures. How then might these elusive but necessary components of life have arisen on the young Earth?

One scenario that may have led to the formation of DNA was beautifully described by Nobel laureate Christian de Duve in his 1995 book *Vital Dust*. De Duve notes that amino acids either would have been brought down to the surface of the young Earth by comets and asteroids from space or would have been created on the planet's surface by chemical reactions. De Duve paints the following picture of our planet more than 4 billion years ago.

> Brought down by rainfall and by comets and meteorites, the products of these chemical re-shufflings progressively formed an organic blanket around the lifeless surface of our newly condensed planet. Everything became coated with a carbon-rich film, openly exposed to the impacts of falling celestial bodies, the shocks of earthquakes, the fumes and fires of volcanic eruptions, the vagaries of climate, and daily baths of ultraviolet radiation. Rivers and streams carried these materials down to the sea where the materials accumulated until the primitive oceans reached the consistency of hot dilute soup, to quote a famous line from the British geneticist J.B.S. Haldane. In rapidly evaporating inland lakes and lagoons, the soup thickened to a rich puree. In some areas, it seeped into the inner depths of the Earth, violently gushing back as steamy geysers and boiling underwater jets. All these exposures and churning induced many chemical modifications and interactions among the original components showered from the skies.

De Duve maintains a long-held belief that the progression from abiotic to biotic was as follows: Amino acids formed in space and on Earth; these next combined to form primitive proteins, which then somehow united to form early life. The crucial step is the formation of proteins, themselves composed of amino acids joined together by chemical bonding. Why? Because formation of the critical building blocks, the nucleic acids, would require enzymes to catalyze the necessary chemical reactions. Most chemical reactions are reversible; sodium and chlorine, for example, combine to form salt under certain conditions and tear apart (or dissolve) under others. Enzymes mediate chemical reactions, which are necessary to join many complex protein pieces together into larger units such as amino acids, and all biological enzymes are proteins.

The need to have proteins already present in order to assemble the molecules whose job it is to assemble proteins in the first place has seemed an intractable "chicken and egg" problem. But recently an elegant solution to this apparent paradox has been proposed. What if one of the nucleic acids—in this case, RNA—could act both as the factory building proteins and as the catalyst necessary to favor the important chemical reactions? According to this new model, the early pathway to life may have seen the formation of RNA *prior to* the formation of protein. In this view, RNA itself acted as the enzymatic catalyst necessary for any further progression toward the ultimate and quintessential component of life, DNA. Francis Crick first suggested this in 1957. Information flows only from the nucleic acids to proteins, he noted, never in the opposite direction, and thus the nucleic acids had to precede protein formation. This point of view was confirmed by the Nobel Prize-winning discovery of Thomas Cech and Sidney Altman that RNA can indeed act as the enzyme necessary for catalytic activity. These RNA enzymes, which were named ribozymes, led to the concept of the "RNA world," where RNA molecules on the early Earth carried out the steps necessary to produce the building blocks of true life, preceding the formation of the first true DNA.

Once RNA has been synthesized, the path toward life is open because RNA can eventually produce DNA. Thus, how the first RNA came into existence—under what conditions, and in what environments—became the central problem confronting chemists. As de Duve notes, "We must now face the chemical problems raised by the abiotic synthesis of an RNA molecule. These problems are far from trivial." The abiotic synthesis of RNA remains the most enigmatic step in the evolution of the first life, for no one has yet succeeded in creating RNA.

Once RNA was created, the leap from RNA to DNA would have been more straightforward. RNA serves as a template for DNA. Yet many mysteries remain: Did it happen once or many times? Was this most vital ingredient of life created over and over and each time snuffed out by another gigantic meteor impact? Or did this essential breakthrough happen just once on Earth and then spread across the planet with its infectious, replicating behavior?

This model of life's origin—from macromolecules to RNA to an "RNA world" to DNA—has not gone unchallenged. Another possibility is that the cradle of life was clay or pyrite crystals. The faces of these flat minerals and crystals could have presented microscopic regions where early organic molecules accumulated. This model suggests the following progression: from clay (mineral) crystals to crystal growth, followed by an "organic takeover" (where purely inorganic molecules are replaced by carbon-based molecules), allowing the formation of organic macromolecules that in turn led to DNA and cells. As envisioned by R. Cairnes, the earliest life would have had several characteristics: It could evolve; it was "low-tech," with few genes (sites on the DNA molecule that code for the formation of specific proteins) and little specialization; and it was made of geochemicals, arising from condensation reactions on solid surfaces, from either pyrite or iron sulfide membranes.

Both of these ideas about the first development of life have at their crux the need to bring various chemical components together somehow and then, from these aggregates, assemble very complex molecules. In the RNA model, the various chemicals assemble in liquid; in the second model, a mineral template becomes an assembly site. There is as yet no consensus on which of these alternatives is correct—or even on whether they are the only alternatives.

How Long Did It Take?

The fossil record tells us that abundant organisms capable of responding to light and able to build mounds existed on Earth 3.5 billion years ago, as evidenced in rocks from the Warrawoona region of Australia. Yet we know that only 300 million years before that—somewhere around 3.8 billion years ago—Earth was still being bombarded by giant asteroids and comets during the "heavy bombardment" phase. This seems like an awfully short period of time for the first life to evolve. Stanley Miller (the chemist who, with Harold Urey, showed in the 1950s that amino acids could be made in a test tube) has in the 1990s derived an estimate on how long it took to go from inorganic chemicals to life. Miller thinks the transition from "prebiotic soup" to cyanobacteria (the microbes we find today sliming swamps and ponds) may have taken as little as 10 million years.

Miller based his conclusion on three lines of evidence: the rate of plausible chemical reactions leading to the formation of the building blocks of life; the relative stability of these building blocks, once made (the number of years they remain intact before decomposition); and the rates of new gene formation through "amplification" in modern bacteria.

The first of these, the rate of amino acid synthesis, is very fast—from minutes to tens of years at the most. Once formed, most organic compounds (such as sugars, fatty acids, peptides, and even RNA and DNA) can last from tens of years to thousands of years. Thus these are not rate-limiting steps; putting the pieces together was what took time. Miller sees three bottlenecks: (1) the origin of replicating systems—essentially the formation of RNA and then that of DNA, which could duplicate itself; (2) the emergence of protein biosynthesis, or the ability of the RNA molecule to begin synthesizing proteins, the actual material of cells; and (3) the evolutionary development of the various, essential cell operations, such as DNA replication, production of ATP (the energy source within cells), and other basic metabolic pathways. In a 1996 article written with Antonio Lazcano, Miller argues that the time necessary to go from soup to bugs may have been far less than 10 million years. Making life may be a rapid operation—a key observation supporting our contention that life may be very common in the Universe.

WHERE DID IT HAPPEN?

Almost as controversial as the "how" and the "how long" of attaining life is the "where." In what physical environment did the first life on Earth originate? Answering the question of *where* is also an important aspect of assessing the possibility and frequency of life on other planets.

The first, most famous, and longest-accepted model was proposed by Charles Darwin, who in a letter to a friend suggested that life began in some sort of "shallow, sun-warmed pond." This type of environment, be it of fresh water or perhaps in a tide pool at the edge of the sea, still remains a viable candidate. Other scientists early in the twentieth century, such as J.B.S. Haldane and A. Oparin, agreed with Darwin and expanded on this idea. They independently hypothesized that the early Earth had a "reducing" atmosphere (one that produces chemical reactions the opposite of oxidation; in such an environment, iron would never rust). The atmosphere at that time may have been filled with methane and ammonia, forming (because it was filled with the chemicals necessary to create amino acids) an ideal "primordial soup" from which the first life appeared in some shallow body of water. Until the 1950s and into the 1960s, it was thus believed that the early Earth's atmosphere would have allowed commonplace inorganic synthesis of the organic building blocks called amino acids by the simple addition of water and energy, as shown in the famous experiments of Miller and Urey in 1952. All that was needed was a convenient place for all the various chemicals to accumulate. The best place for this seemed to be a fetid pond or a wave-lapped tide pool at the edge of a shallow, warm sea.

Yet as we learn more about the nature of our planet's early environments, tranquil ponds or tide pools seem less and less likely to be plausible sites for the first life—or even to have existed at all on the surface of the early Earth. What Darwin could not appreciate in his time (nor could Haldane and Oparin, for that matter) was that the mechanisms leading to the accretion of Earth (and of other terrestrial planets) produced a world that, early in its history, was harsh and poisonous, a place very far removed from the idyllic tide pool or pond envisioned in the nineteenth and early twentieth centuries. In fact, we now have a very different view of the nature of the early Earth's atmosphere and chemistry. It is widely believed among planetary scientists that carbon dioxide, not ammonia and methane, dominated the earliest atmosphere and that the overall conditions may not have favored the widespread synthesis of organic molecules on Earth's surface. It seems more reasonable that the rain of asteroids and comets delivered these compounds essential for life.

But if not in a pond or tide pool, where could these components have come together to produce life? Here is an alternative view from microbiologist Norman Pace, one of the great pioneering microbiologists interested in life's evolution:

> We can now imagine, based on solid results, a fairly credible scenario for the terrestrial events that set the stage for the origin of life. It seems fairly clear, now, that the early earth was, in essence, a molten ball with an atmosphere of high-pressure steam, carbon dioxide, nitrogen, and other products of volcanic emissions from the differentiating planet. It seems unlikely that any landmass would have reached above the waves (of a global ocean) to form the "tide pools" invoked by some theories for the origin of life.

Pace looks for an entirely different setting—one of heat and pressure, such as in the deep-sea volcanic vents.

The "where" of life's origination is obviously controversial, and as pointed out by University of Washington astronomer Guillermo Gonzalez, the favored habitats appear to depend on a given scientist's discipline. In his delightful 1998 essay "Extraterrestrials: A Modern View," Gonzalez noted,

> The kind of origin of life theory a scientist holds to seems to depend on his/her field of specialty: oceanographers like to think it began in a deep sea thermal vent, biochemists like Stanley Miller prefer a warm tidal pool on the Earth's surface, astronomers insist that comets played an essential role by delivering complex molecules, and scientists who write science fiction part time imagine that the Earth was "seeded" by interstellar microbes. The fact that life appeared soon after the termination of the heavy bombardment about 3.8 billion years ago tells little about the probability of the origin of life—it could have been a unique event requiring extraordinary conditions. However, there are a few very basic ingredients that are required by any conceivable kind of life, overactive imaginations notwithstanding.

Our vision of the "cradle of life" has obviously changed since Darwin's time. How do scientists now envision planet Earth at the time of life's first appearance? Even around 4 billion years ago, about 500 million years after initial accretions, Earth would have seemed a very foreign world to us. For instance, there was little land area, because there were few or no continents. Volcanism and the eruption of lava from the interior of the planet, however, were far more common than today. The deep-sea ridges, places where new oceanic crust is created on the sea floor, are estimated to have been three to five times longer than they are today, and hydrothermal activity along these ridges may have been as much as eight times greater than in the present-day world. All of this suggests a very energy-rich, volcanic world, with huge amounts of deep-earth chemicals and compounds spewing forth in the oceanic environment. The chemistry of seawater would have been enormously different than it is now. The ocean was what we would call "reducing" (in contrast to the present-day oxidizing oceans) because there was no free oxygen dissolved in the seawater. The temperatures of the oceans may have been far higher than today, ranging from warm to hot—perhaps hot enough to scald us if we were there. Finally, there may have been 100 to 1000 times as much carbon dioxide in the atmosphere as there is today.

The extremophiles may thus yield the most important clue uncovered to date. Darwin and de Duve imply that life originated on Earth's surface (although de Duve hedges a bit on this question with his comment that environments within the planet may be involved as well). Yet most views of Earth's surface at the time of the first formation of life paint a very bleak picture. Lethal levels of ultraviolet radiation polluted the surface, and the impacts of giant comets with Earth periodically vaporized the planet's oceans. The boiling of the seas would have repeatedly sterilized Earth's surface. But what about *beneath* the surface, in the subterranean regions now inhabited by the extremophilic archaeans and bacteria? These deep Hadean environments may have served as bomb shelter-like refuges, protecting deep extremophiles from the fury at the planet's surface. Could the deep subsurface have served not only as refuge but also as cradle of life early in Earth history? New analyses of the "Tree of Life," or phylogenetic history of life on our planet, support this possibility. But before we examine the Tree of Life and what it tells us, we need to consider one more possible origin of life on Earth.

PLANETARY CROSS-TALK

There is another reason why life—at least microbial life—may be widely distributed: Planets may commonly be seeded by life from other, nearby planets. This may have happened on Earth; perhaps life arose on Mars or Venus and then seeded our Earth. If microorganisms, the primitive but nearly indestructible creatures at the low end of the cosmic IQ scale, exist on a given world, they must *inevitably* travel to its immediate neighbors. There is a natural "interplanetary transportation system" that distributes rocks between nearby planets. These rocks serve as natural spacecraft that are capable of carrying unwitting microbial stowaways from the surface of one planet, across hundreds of millions of miles of space, to neighboring planets. This process has nothing to do with the inclinations or technology of the inhabitants. It is an unavoidable act of nature. Each year, Earth is impacted by half a dozen 1-pound or larger rocks from Mars. These rocks were blasted off Mars by large impacts and found their way to Earth-crossing orbits, where they eventually collided with Earth. Nearly 10% of the rocks blasted into space by Mars end up on Earth. All planets are impacted by interplanetary objects large and small over their entire lifetimes, and the larger impacts actually eject rocks into space and into orbit about the sun.

A glance at the full Moon with binoculars shows long streaks, or rays, radiating from the crater Tycho, located near the bottom of the Moon as seen by observers in the Northern hemisphere. The rays are produced by the fallback of impact debris (rocks) ejected from the crater, which is 100 kilometers in diameter. The rays can be traced nearly across the full observable side of the Moon, and such long "airborne" flight is evidence that some ejecta were accelerated to near-orbital speed. Debris ejected to speeds higher than the escape speed (2.2 kilometers per second) did not fall back but flew into space. It has long been appreciated that material could be ejected from the Moon by impacts, but only in the past decade have we realized that whole rocks greater than 10 kilograms in mass could be ejected from terrestrial planets and not be severely modified by the process. It was formerly believed that the launch process would shock-melt or at least severely heat ejected material. There was little expectation that rocks capable of carrying living microbes from planet to planet would survive the great violence of the launch. The discovery of lunar rocks in Antarctica showed that this is possible.

There is also a rare class of meteorites called SNCs, or "Martian meteorites," that are widely believed to be from Mars. The first suggestion that these odd meteorites might be Martian was greeted with considerable skepticism. The discovery of lunar meteorites changed this by proving that there actually was an adequate natural launch mechanism. The lunar meteorites could be positively identified, because rocks retrieved by the Apollo program showed that lunar samples have distinctive properties that distinguish them from terrestrial rocks and normal meteorites derived from asteroids. Positive linking of the SNC meteorites with a Martian origin was a more complex process. It included showing that noble gas trapped in glass in the meteorite served as a telltale fingerprint that matched the composition of the Martian atmosphere, as measured by the Viking spacecraft that landed on Mars in 1976. The general properties of the SNC meteorites revealed that they were basalts formed on a large, geologically active body that was definitely neither Earth nor the Moon. Because the atmosphere of Venus is too thick and its surface too young, Venus was also ruled out as a source.

The astounding discovery that meteorites from the Moon and Mars reach Earth has profound implications for the transport of life from one planet to another. Over Earth's lifetime, billions of football-size Martian rocks have landed on its surface. Some were sterilized by the heat of launch or by their long transit time in space, but some were not. Some Martian ejecta are only gently heated and reach Earth in only a few months. This interplanetary shuttle is capable of carrying microbial life from planet to planet. Like plants releasing seeds into the wind, or palms dropping coconuts into the ocean, planets with life could seed their neighbors. Perhaps, then, life on nearby terrestrial planets might have common origins. The seeding process would be most efficient for planets that have small velocities of escape and thin atmospheres. In this regard, Mars is a better prospect than Earth or Venus. That is why it has been suggested that terrestrial life may have been seeded by Mars.

What about the transfer of microbes between stellar systems? Although microbes are killed by radiation in space some bacteria or viruses embedded in dust grains might be shielded sufficiently to survive. If so, they might possibly "seed" regions of a galaxy through the process known as Panspermia, as suggested by Fred Hoyle and his collaborators in the early 1980s.

Once any planet in a particular planetary system is "infected" with life, natural processes may spread that life to other systems. Of course, this process can work only on organisms that can withstand the raw vacuum of outer space. Animal life cannot spread in this fashion.

THE TREE OF LIFE AND THE ORIGIN OF THE EXTREMOPHILES

Once originated (or infected from elsewhere), life on Earth evolved quickly. Geneticists have proposed several possible scenarios for this first unfolding of Earth's biota.

The first great surprise delivered by the Archaea, the extremophilic microbes on Earth, was that they could live in such extreme environments. The second and equally dramatic discovery was that archaeans are among the most ancient of extant organisms on Earth, and display some characters believed to be primitive. Studies of the genes of bacteria and archaeans (using powerful techniques of molecular biology) have shown that both appear to be near the very base of the so-called Tree of Life (also known as the "Woese rooted tree of life" after its discoverer, geneticist C.R. Woese; see Figure 4.1).

The Tree of Life is really a model of life's evolution into the major categories of organisms existing today, and as such it is built simply of a series of hypotheses in which we have greater or lesser degrees of confidence. Studies that compare gene sequences in various organisms give us a theoretical map

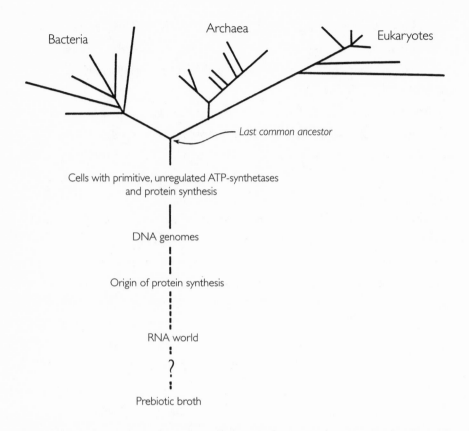

Figure 4.1 *The origin and early evolution of cells beginning from an RNA world (see page 65). The branching order depicted in the upper part of the tree is from Woese et al., 1990. The distances separating the evolutionary events in the tree trunk are not drawn to scale. (Modified from Lazcano et al., 1992.)*

of evolutionary history. According to these new studies, there is little more "primitive" still living on Earth than hyperthermophilic microbes (although we should note that *primitive* is used here in the sense of being first—these are still very sophisticated cells, superbly adapted for their mode of life). On the basis of the various genetic studies conducted to date, the archaeans seem to share more characteristics and genes with the supposed primordial organism (the putative common ancestor of all life) than any other organism living on

Earth. But they have still undergone more than 3.8 billion years of evolution and hence may be very different from that first strain of life.

Making sense (and understanding the order) of life's diversity is the specialty of systematic biology. Early systematists simply classified through similarities and differences of body parts. Now we classify by evolutionary *history*, not simple resemblance. Although the presence of shared traits is often a powerful clue to an evolutionary pathway, it can be quite misleading. Insects, bats, birds, and pterosaurs all fly (or flew), yet they are only distantly related. A far more powerful method of classification is through finding the presence not only of shared *characters* (which can be anatomical details, such as the presence of a backbone) but also of shared *derived characters*—indications that these characters have been shaped by evolutionary forces and then passed on. This particular methodology, coupled with new advances in DNA sequencing, has led to a breakthrough in understanding life's categories and evolutionary history. The analysis of molecular sequences from living organisms has provided a rough "map" of life's evolution. Portrayed in graphical form, this map becomes the "tree" mentioned above. The greater the number of differences between the genes, the more evolutionarily separated the groups are. It was this technique that illuminated the existence of the three "domains," Archaea, Bacteria, and Eucarya, and showed that these three are the most ancient and fundamental branches of the Tree of Life still present on the planet. This analysis also showed that bacteria and archaea are distinct, even though they have some similar attributes, such as a cell without an internal nucleus.

At first it would seem unlikely that the gene sequences preserved in still-living organisms could yield *any* sort of accurate key to the past, especially a past of such antiquity, for the sequencing is an attempt to unravel life's first diversification, which took place more than 3 billion years ago. And yet, at least in some molecules, evolutionary change has been exceedingly slow. The most convenient places to study rates of evolutionary change within cells are small subunits of RNA extracted from ribosomes, tiny organelles found within all cells; these have been the Rosetta Stone that has enabled us to construct a new view of the ancient past using only the recent past as our evidence.

Much of this work has been accomplished in the 1990s and contradicts long-held beliefs about the phylogeny, or evolutionary pathway, of life. It shows that the divisions between the domains are extremely ancient. But by far the most intriguing result is evidence that the most ancient of extant archaeans and bacteria are heat-loving extremophiles—just the types of microbes we find in extreme environments on Earth today. They are also among the most slowly evolving organisms. This discovery indicates either that the earliest life on Earth was some sort of extremophile or that extremophiles have best survived the numerous near sterilizations of the early Earth. The implications are of the utmost importance to those trying to estimate how common life is on other planets. It appears that life may have first arisen on Earth under conditions of high temperature and pressure, either under water or deep in Earth's crust. As we noted before, it may be that life can originate in settings far harsher—and thus far more common in the Universe—than we ever dreamed.

The view that extremophilic microbes offer clues to the environment of life's first formation on Earth is relatively recent. In a scientific paper published in 1985, John Baross and S. Hoffman argued that life first arose in the deep-sea hydrothermal vent systems. At that time the extremophilic microbes in such settings had only recently been discovered. Yet it seemed to Baross and Hoffman that the early hydrothermal sites, as well as Earth's deep crustal regions, offered both the chemicals and energy necessary to *form* the first life and the refuges it would need to remain alive during the violent early phases of Earth history. After all, compared to heavy asteroid bombardment, the deep-ocean floor near hydrothermal vent systems, violent as they are, would constitute a relatively stable environment—perhaps the only environment on Earth suitable for life's first formation and first flowering. The scientific community largely dismissed this new hypothesis when it was first advanced, for the ubiquity and variety of extremophile microbes in such settings were not yet known. But as the hydrothermal vent hypothesis garnered support from evolutionary Tree of Life studies, many came to regard the deep-vent setting as the strongest candidate for the site of life's origin.

Several key properties of hydrothermal vents make this hypothesis attractive. First, hydrothermal vents contain regions where temperature, acidity, and chemical content are favorable for life. They also contain a suite of ingredients that make up the recipe for life, such as organic compounds, hydrogen, oxygen, and abundant energy in appropriate energy gradients. They offer reactive surfaces—places on rock substrates that might act as templates for early protein formation. And most important, perhaps, they exist today and allow us to test the plausibility of this hypothesis.

The most convincing interpretation linking the origin of life on Earth with hydrothermal systems has come from astrobiologist Everett Shock and colleagues at Washington University. Shock notes that unlike the oceans, the early *atmosphere* was probably not a reducing environment. (This assumption is contrary to that of others, who believe that the atmosphere may have remained reducing for an extended period of time, thus providing environments where organic compounds may have formed in a manner after the famous Miller–Urey synthesis experiments of the early 1950s.) Shock argues that in the absence of a reducing atmosphere, the synthesis of organic compounds such as methane and ammonia—necessary building blocks for life—would have been impossible on Earth's surface. Instead, the first reactions to build organic compounds would have been conversion of the common gas carbon dioxide and perhaps carbon monoxide to organic compounds. This is a radically different scenario from the idea that simply zapping the early ocean with lightning (as envisioned by Miller and Urey) would create organic compounds, which would then somehow fuse together to form the first life.

Shock also argues that the surface of the early Earth would have been a most inhospitable place for life's origin, because it would have been bombarded by ultraviolet radiation and cosmic debris. Like John Baross, Jody Deming, and others, Shock comes down strongly on the side of submarine hydrothermal systems as the cradle of life. These systems would have provided the combination of high temperatures and chemical environment (reducing conditions) necessary for converting carbon dioxide into organic material. The chemical energy for this synthesis would have come from the mixing of highly reduced fluids rich in the noxious gas hydrogen sulfide with less-reduced seawater in the very gullets of the hydrothermal vent systems. Mixing of these two different solutions would provide energy—chemical energy. This same energy source is the foundation of modern deep-sea vent communities of organisms. In such a world, the earliest metabolic systems of life would have been "chemoautotrophic"—existing not by carrying on photosynthesis or by eating other creatures but, rather, deriving energy from chemical reactions in seawater.

Much of the more recent debate about the origin of life has centered on whether it occurred in an environment that was truly "hot" (above the boiling point of water), such as is found well within the volcanic hydrothermal vents, or in an environment merely "warm." If the first life used RNA rather than DNA for genetic information, it seems unlikely that the "hot" environment could have been life's cradle, because RNA is much less stable than DNA under conditions of heating. RNA would probably be unable to develop or evolve above 100°C, and such temperatures are routinely found within the hydrothermal vent systems. It may be that, contrary to the interpretation of those who use the Tree of Life as primary evidence, the earliest life originated as mesophilic (warm-loving) rather than thermophilic (heat-loving) microbes. Under this scenario, the true heat-loving forms evolved from the warm-loving forms and may have been the only survivors of the cometary holocaust, all "mesophiles" of the time having been overheated into extinction.

This debate will not soon be over. There is no way to know to what extent the ancient ancestors of the living extremophiles resembled those microbes still on our planet. John Baross has pointed out that the period between 4.0 and 3.5 billion years ago may have been a time of extensive evolutionary "experimentation," with only one evolutionary lineage becoming the source of extant organisms. The tree accepted until 1997 may record the lone survivors of that long ago time, rather than the true first ancestor of all life on this planet. In such a case, the basal "trunk" of the tree is no more than a branch extending from a much more deeply rooted tree whose other, more ancient branches have been stripped from Earth through extinction.

By 1998 the "Tree of Life" had again changed in appearance (see Figure 4.3). The details of its upper branches remained about the same as in the 1997 tree, but the shape of its base began to look different. This reorganization was based on new DNA-sequencing results from a microbe called *Aquifex*, a thermophile that lives in hot springs in Yellowstone Park. *Aquifex* had its entire assemblage of genes decoded. To the surprise of many (who hoped that it was very similar to the most primitive of all life), *Aquifex* turned out to have a gene assemblage not so different from that of many other, nonextremophilic microbes. In fact, this thermophile differed in only a single gene sequence from microbes that can live at normal temperatures. The implication is that microbes that belong to widely divergent biological groups—(including, perhaps, different *domains*) seem to have been able to exchange entire blocks of genes—a process called gene swapping—very early in their history. Gene swapping, or lateral gene transfer, must have been a radical and common form of genetic exchange.

If gene swapping was so easily managed by the first groups of life on Earth, it might help explain why all life (at least all life on Earth) uses the same genetic code. Carl Woese supposes that all three domains—Archaea, Bacteria, and Eucarya—emerged from the shared pool of genes that commonly were transferred from organism to organism by the process of lateral gene transfer. Innovations cropping up in any individual were quickly shared and assimilated by others in the gene pool. Eventually, as more and more complex proteins came into existence, and as the combinations of genes coding for protein formation became more and more complex, the three domains emerged.

The standard view, that Bacteria and Archaea are the two oldest groups, and that the eucaryans descended from one of them, now has two alternatives: that all three groups emerged from a common gene "pool" or that there once was a fourth, even more primitive domain that gave rise to the rest but is now extinct (see Figure 4.2).

Whatever and wherever its source, life was rooted and probably pervasive on the planet Earth by 3.5 billion years ago. Evolution was at work, and

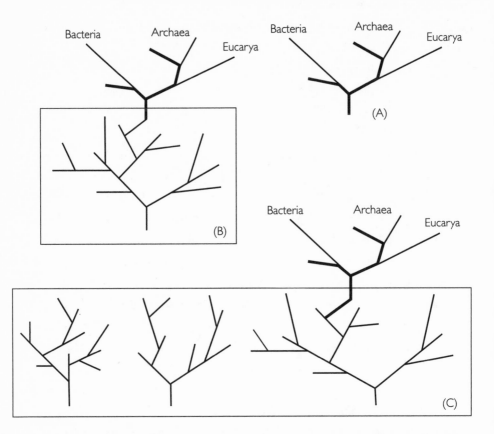

Figure 4.2   *Three alternatives for the evolution of life on Earth and its "Tree of Life." In (A), the three do-mains of life are seen to originate from a single common ancestor. This is the widely accepted "Tree" found in most texts today. In (B), the tree seen in (A) sits atop an older and currently unrecognized series of branches. In this interpretation, DNA life as we know it had a long prehistory with no preserved record. In (C), life is composed of several distinct types that independently evolved on the early Earth, with only one (DNA life) sur-viving.*

a host of new species proliferated as life began to exploit new food, new habi-tat, and new opportunities. The possible ways in which life may have first formed and the speed of its formation suggest that life may not be a unique property of this planet. Perhaps it can be found on any planet or moon with heat, hydrogen, and a little water in a rocky crust. Such environments are common in our solar system and probably in other parts of the galaxy and Universe, so life itself may indeed be widespread. The lesson of Earth is that

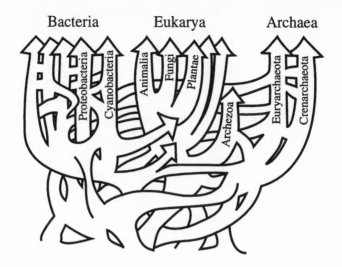

Figure 4.3   *A reticulated tree representing Life's history. Adapted from Doolittle 1999.*

not only can it live in extreme environments, it can probably form in such places as well. Life—*but not animal life.* How that further step to animal life occurred on Earth—and whether we can use this essential step in Earth's history as a means of modeling the evolution and formation of analogs to animal life on other planets—is the subject of the next chapter.

How to
Build Animals

Surely, the mitochondrion that first entered another cell
was not thinking about the future benefits of cooperation
and integration; it was merely trying to make its own living
in a tough Darwinian world.

—Stephen Jay Gould

On a perfect planet such as might be acceptable to a
physicist, one might predict that from its origin the
diversity of life would grow exponentially until the carrying
capacity, however defined, was reached. The fossil record
of the Earth, however, tells a very different story.

—Simon Conway Morris

In earlier chapters we've seen that life can exist in environments previously
thought too rigorous or too extreme to support living cells. We have also
seen that not only can life *exist* in these extreme environments, but on
Earth, at least, it may have originated in them as well. The implication of
these recent findings is that because microbial life can survive and perhaps
originate in extreme environments, it may be widespread in the Universe—

and even on other planets in our solar system. Yet what of higher life forms? Will multicellular animals and plants be as common as bacteria on other planets? In this chapter we will examine these questions by looking at how higher life forms came to be on planet Earth and asking, as we did in the case of the extremophiles, whether this particular history can lead to generalizations or insights concerning the frequency of animal life beyond Earth.

An Ancient Dichotomy

The gulf between the complexity of a bacterium and the complexity of even the simplest multicellular animal, such as a flatworm like *Planaria*, is immense. The number of genes in a bacterium can be measured in the thousands, whereas the genes in a large animal number in the millions. To illustrate this, we can liken a bacterium to a simple toy wooden sailboat. With only three or four very tough parts, the toy boat is virtually indestructible, just as a bacterium is impervious to most environmental stress. The flatworm, by contrast, is like an ocean liner: immensely larger, more complex, and the product of countless technological achievements. The sailboat does not need complex fuel; it uses wind as its energy source, just as an autotrophic bacterium (one that does not require organic nutrients) can take the simplest sources, such as hydrogen and carbon dioxide, and manufacture its own organic material. A planarian must find and ingest complex food, and it needs a wide range of nutrients and inorganic materials to live, just as an ocean liner must be supplied with complex fuel and devotes much of its internal machinery to converting fuel to motion and energy. Let us pursue this simple analogy further and bring in the time component. Because their technology is so simple, toy sailboats have been built by humans for thousand of years. Ocean liners, on the other hand, are a product of this century. They had to await the development of complex smelted metals, steam or internal combustion engines, electronics, and all the rest. They cannot be built simply, nor could they be built until each of their various components was first invented and perfected. Sailboats (toy or otherwise) have been on Earth a long time. Not so ocean liners—or even the simplest of animals.

There is a final parallel we can draw. Like all objects built by the hand of humans, our toy sailboat will eventually be destroyed: It will perhaps lose first its cloth sail and then its mast; eventually the wood of the hull will rot. But until then it is virtually unsinkable, just as the microbes of this planet not only can withstand a much larger range of conditions than any animal but seem to resist extinction much longer as well. Our ocean liner, on the other hand, is a very different "animal." One of the first of this century, of course, was named *Titanic.*

The animals now on our planet are distinct from the domain Bacteria and from the other bacterium-like domain, Archaea. We animals belong to a third branch, the domain Eucarya (all three domains, however, share a common ancestor).

As we noted in an earlier chapter, living organisms were long classified into two great "kingdoms" composed of animals and plants. Eventually that number was increased to five, made up of the animals, plants, bacteria, fungi, and protists as noted in the last chapter. Modern classifications now break life into three more basic groupings termed domains: Eubacteria (or simply Bacteria), Archaea (whose members are united by exhibiting a cell type known as prokaryotic), and Eucarya (composed of everything else, including the former standard-bearers, animals and plants). The major subdivision is thus no longer along classical lines (which, itself, was based on the well-known differences between animals and plants) but is based instead on major differences in cell architecture and genetic content between these groups.

The archaeans and the bacteria (which we can hereafter refer to as prokaryotes) have no internal nucleus and no membrane-bounded organelles. Their genetic information is contained in a single strand of DNA that is embedded in the cell's interior cytoplasm and thus separated from the external environment only by the outer cell wall. The prokaryotes reproduce largely by asexual means. They grow rapidly and divide frequently. The eucaryans are so genetically distinct from the bacteria and the archaeans that they are easily differentiated as a third separate domain. Eucaryans have a very different internal anatomy and organization as well. They possess an internal nucleus and other internal cell compartments (known as organelles), such as the mitochondria that produce energy.

Let's clarify a few terms before we continue. The term *clade* refers to a group of organisms that share a common ancestor more recently than some other group of organisms. The archaeans, bacteria, and eucaryans make up separate clades. The term *grade*, on the other hand, refers to a level of organization. For instance, mammals and birds are both warm-blooded and thus share the "warm-blooded" grade, even though the two groups are members of different clades. We use the terms *prokaryote* (and *prokaryotic*) and *eukaryote* (and *eukaryotic*) to denote two different grades of evolution. All bacteria and archaeans belong to the prokaryotic grade, even though they represent separate clades. Eucaryans belong to the eukaryotic grade.

Yet there is a far more fundamental distinction between these groups than simple architecture or differences in their genetic code. These three groups have evolved very different strategies for coping with environmental challenges. Archaeans and bacteria tend to solve their problems by using chemistry: They have evolved innumerable metabolic solutions to Earth's environmental challenges over time, but they have changed their morphology very little in so doing. Perhaps because of this, the archaeans and bacteria have attained very limited morphological diversification compared to the immense number of species that have evolved in the domain Eucarya. Most of them have retained the single-cell body plan. What they *have* done, with sweeping success, is evolve a wide variety of metabolic specializations and thus find biochemical and metabolic solutions to environmental challenges. When archaeans and bacteria encounter an environment not to their liking, they literally try to change the chemical nature of their surroundings.

In contrast to the mainly single-celled archaeans and bacteria, most eucaryans have taken the opposite tack: They respond to challenges by changing or creating new body parts. Theirs has been a morphological, rather than a metabolic, approach. One consequence of this mode of life has been an increase in size. The eucaryans evolved forms with an internal nucleus and other internal cell organelles; this led to larger bodies. And they also mastered the art of incorporating many cells per individual.

The earliest known fossil record of life, from rocks dated to about 3.5 billion years ago, appears to be of either archaeans or bacteria, which suggests that one of these two groups may include the earliest truly living organisms to have evolved on Earth. These first fossils are filamentous, and they closely resemble the extant filamentous bacteria known as cyanobacteria. The continued existence of these forms suggests that these ancient prokaryotes achieved early a level of success that has not required major subsequent morphological fine-tuning. But couldn't the interiors of these cells, now separated by 3.5 or more billions of years, be radically different? Possibly, but not likely. Given a time machine capable of transporting us to Earth's first cradles of life, we would in all probability bring back microbes morphologically, chemically, and perhaps even genetically indistinguishable (or nearly so) from extant forms. Scientists have come to this conclusion through their efforts at decoding the sequence and function of genes in modern bacteria. Each gene has a function or series of functions, and because many living bacteria are found in environments similar to those of the ancient Earth, it can be assumed that to survive, the ancient bacteria had to have very similar genes. The genetic code of many microbes is still very basic—and probably not much different from that of types living over 3 billion years ago. The bacteria and archaeans appear to be highly conserved; that is, they are true living fossils. And in addition to being very old, they have been very successful. They remain the most abundant living forms on Earth. There can easily be as many bacteria in a drop of water as there are humans on all of Earth. We inhabit, and always have, the "Age of Bacteria."

The evolutionary history of the bacteria and archaeans is thus one of little morphological change over the past 4 billion years. The evolutionary history of the third great domain, Eucarya, was decidedly different (see Figure 5.1). A few of the eucaryans retained their primitive, almost bacterium-like

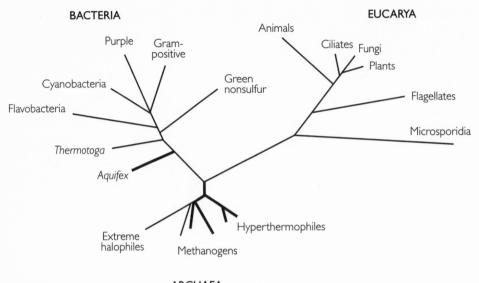

Figure 5.1 *An "unrooted" Tree of Life. The three major domains of life (Archaea, Bacteria, and Eucarya) all extend out from a central point. The principal taxonomic categories of each domain are shown as branches.*

ways, and some of these still exist on Earth today. The remainder, however, achieved one of life's most remarkable transformations by creating a new cell type, the eukaryotic cell, whose greatest innovation is the presence of an internal nucleus. It was from this group that animal life eventually evolved. Let us now examine the differences between the prokaryotic and eukaryotic grades. This distinction has great relevance to our story, for it appears that attaining the eukaryotic grade was the single most important step in the evolutionary process that culminated in animals on planet Earth.

In prokaryotic cells, the most important barrier between the cell and the external world is the cell wall. In eukaryotic cells, many barriers exist, including the wall to the nucleus, the cell wall, and, among the multicellular varieties, the epithelium (the outer skin). The eukaryotes have followed the path of compartmentalizing cell functions in membrane-bounded organelles, such as the nucleus, the mitochondria, chloroplasts, and others. Because of this, the morphological differences between prokaryotes and eukaryotes are profound. Yet other, nonmorphological distinctions exist as well, and these too have affected the evolutionary histories of these groups.

The most obvious difference is related to the varying degrees of multicellular organization achieved by prokaryotes and eukaryotes. Prokaryotes have only rarely attained either larger size or "metazoan" levels of organization (many cells in a single organism). Nevertheless, the multicellular forms that *have* evolved have played a starring role in Earth's history. The most important of these are stromatolites (the term means "stone mattress") composed of bedded masses of photosynthetic bacteria. Even when multicellularity has been achieved by prokaryotes, intercellular coordination of tasks has been minimal. Eukaryotes, on the other hand, have repeatedly evolved multicellular forms.

These two strategies have had a marked effect on the evolutionary histories of the two groups. As we have said, some species of bacteria on Earth today seem indistinguishable from fossil forms found in rocks more than 3 billion years old. In contrast, the majority of eukaryotic species with a fossil record (and therefore hard parts) seem to persist for periods of only 5 million years or less. Sexual reproduction and many episodes of evolutionary radiation and extinction resulting in morphological change characterize the majority of metazoan (multicellular) eukaryotes. Prokaryotes, on the other hand, have seemingly adapted a strategy that protects them from extinction but at the same time stifles morphological innovation. These are profoundly different paths. How did this major differentiation of life on Earth come about?

The discovery of the ancient evolutionary divergence between the great prokaryotic sister groups, Bacteria and Archaea, destroyed the long-held belief that these so-called "primitive groups" are closely allied, one being the ancestor of the other. Now many microbiologists believe that both arose from some older, common ancestor as yet unknown. An even more surprising discovery concerned the ancestry of the domain Eucarya. The lineage from which all modern-day animals and plants arose might be as ancient as the two prokaryotic groups. This is not to say that present-day eukaryotic cells are as old as the prokaryotes. Most (but not all) scientists studying this problem believe that the eukaryotic *grade* of cell, with its clearly defined internal nucleus and many advances over the prokaryotic grade, did not appear on Earth until well over a billion and a half years after the first bacteria and archaeans. What this analysis does suggest is that the eukaryotic cell, the basic prerequisite for the formation of complex metazoan life, arose but once—from a group of bacterium-like organisms. The rest was evolutionary history. From that first, complex cell with an internal nucleus arose all subsequent forms that people this diverse domain: plants, fungi, various protist groups (including the flagellates and ciliates, single-celled creatures that inhabit ponds and lakes and are readily observed with a simple microscope), a group of microorganisms called microsporidia—and, ultimately, animals.

THE "NUCLEAR" FAMILY

In summary, we can characterize eukaryotic cells as having seven major characteristics that distinguish them from prokaryotes.

1. In eukaryotes, DNA is contained within a membrane-bounded organelle, the nucleus.

2. Eukaryotes have other enclosed bodies within the cell—the organelles such as mitochondria (which produce energy) and chloroplasts (tiny inclusions that allow photosynthesis).

3. Eukaryotes can perform sexual reproduction.

4. Eukaryotes have flexible cell walls that enable them to engulf other cells through a process known as phagocytosis.

5. Eukaryotes have an internal scaffolding system composed of tiny protein threads that allow them to control the location of their internal organelles. This cytoskeleton also helps eukaryotes replicate their DNA into two identical copies during cell division. This system is both more complicated and more precise than the simple splitting of DNA that occurs in prokaryotic cells.

6. Eukaryotic cells are nearly always much larger than prokaryotes; they usually have cell volumes at least 10,000 times greater than that of the average prokaryotic cell. An internal architecture and salt balance systems far more advanced than those found in prokaryotes that make this larger size possible.

7. Eukaryotes have much more DNA than prokaryotes—usually 1000 times as much. The DNA in the eukaryotic cell is stored in strands, or chromosomes, and is usually present in multiple copies.

Between the first evolution of life and the first cell to sport a eukaryotic grade of organization, perhaps more than 1.5 billion years would pass. Why did it take so long?

The answer partly seems to be that a host of new cell parts and cellular organization had to evolve, and each took time. Perhaps the most important development was a far higher degree of organization within the cell itself than is found in bacteria or archaeans. Much of this organization is a result of the cytoskeletal structures. The gathering of the cell's DNA into an enclosed region, the nucleus, and the compartmentalization of other cell systems into enclosed organelles were radical departures from the prokaryotic design. Some scientists believe that this compartmentalization of the cell interior may also be one of the prerequisites for the development of "complex metazoans"—animals and higher plants.

How did this evolutionary transformation come about? Evolutionary biologist Lynn Margulis and others have argued that the eucaryans evolved their various internal organelles through a process that began with endosymbiosis, wherein one organism lives within another. This discovery, now largely accepted, represents one of the great triumphs of twentieth-century biology. There are many examples of endosymbiosis today; for example, termites and cattle are able to digest the tough cellulose of plant material or wood because they harbor, within their digestive systems, bacteria that contain enzymes capable of breaking down the woody material. These bacteria, however, are unaffected by the host creature's digestive enzymes. Endosymbiosis may have been the first step in the acquisition of the all-important eukaryotic cell organelles, through the following scenario.

In the distant past, some early eucaryans (perhaps already endowed with a nucleus, but still very small and lacking other organelles) may have routinely ingested other prokaryotic cells for food. This in itself was a major advance over the prokaryotes, for it necessitated the evolution of an outer cell wall that could devour, or phagocytize, other cell material—in other words, it made predation possible. Some of these ingested prokaryotic cells, however, were not rapidly digested and hence destroyed within the host cell. Instead, they may have lived there for some time. (Alternatively, these organelles may have invaded the host cell, rather than being captured by it; perhaps they bored in and established parasitic colonies within the larger eukaryotic cell environments.) Eventually, the host cell came to benefit from this association in some way: Prokaryotes, being very efficient chemical factories, may have performed services the host could not carry out for itself, such as energy transformation or even energy acquisition, and metabolic functions. The organelles known as mitochondria (which are involved in energy formation and transformation), plastids (the sites of chlorophyll), and perhaps even flagella (which are used for locomotion) may have evolved in this fashion.

The prime evidence for this hypothesis comes from DNA. Mitochondria and plastids contain their own strands of DNA, which are closer in structure to prokaryotic than to eukaryotic DNA. Mitochondria may have been free-living bacteria that were capable of oxidizing simple carbohydrates into CO_2 and water and liberating energy in the process. There are living bacteria today, such as forms known as purple nonsulfur bacteria, that may be close to the ancestral mitochondrial form. When incorporated into the host cell, these "guests" eventually lose their cell walls and become part of the host. With the addition of the cell organelles, our eukaryote approaches or attains a level of organization that would be familiar to us.

We can now sketch the evolutionary steps necessary to arrive at the eukaryotic grade of organization. We start with a cell membrane enclosing DNA—a simple bag of protoplasm and DNA—and then evolve the ability to phagocytize (or engulf material), evolve a cytoskeleton (which allows us, among other things, to get larger), evolve aerobic respiration, and then bring into our much larger bag various organelles: mitochondria, the nucleus, ribosomes, and so forth.

This last is among the most interesting and controversial aspects of the evolution of the eukaryotic cell type. Few scenarios have been proposed that make evolutionary and adaptive sense, but one intriguing possibility has been described by Dr. Joseph Kirschvink of Cal Tech, who has summarized the problems faced by the evolving eukaryote as follows:

The problems for the eukaryotic host cell are that:

The host must be large enough to engulf other bacteria.

The host cell must be capable of phagocytosis, so that the invaders are put into a membrane-bound vacuole (a small space within the cell), leading to the characteristic double membrane of the mitochondria and chloroplasts.

The cell should have at least a rudimentary cytoskeleton.

The host cell should offer a better, more controlled environment for the symbionts, so that natural selection would favor the association.

The only known bacterium that meets all of these constraints is called *Magnetobacter*, discovered in Germany, which dwarfs most other protists (in size). Each cell of this bacterium makes several thousand organelles called magnetosomes, which are tiny crystals of the mineral magnetite (Fe_3O_4) encased in a membrane bubble—a bubble that forms by phagocytosis. These magnetosomes are held in place in chain-like structures that keep each crystal aligned properly; this can only be done if an intracellular mechanical support structure such as the cytoskeleton exists. *Magnetobacter* has the ability to keep itself in the optimal environment by swimming along the magnetic field lines generated by the Earth's magnetic field. This ability makes it an attractive partner for symbiosis, as many organisms spend a great deal of their metabolic energy staying in the correct environment.

This scenario for the evolution of the eukaryotic cell has two major implications for the timing of higher life on Earth. First, the most probable path for the evolution of magnetotaxis (the ability to align along magnetic fields) and magnetite biomineralization (the formation of minerals by living organisms) is a result of natural selection for iron storage. Anaerobic microbes do not need iron storage mechanisms, as ferrous iron is freely available in solution. But in oxygen-rich environments the iron rusts out the [ferrous] form, which drops out of solution. Hence, magnetotaxis is unlikely to evolve in an anaerobic world, which on Earth ended about 2.5 to 2 billion years ago. The oldest magnetofossil—the fossil remains of bacterial magnetosomes—date to about 2 billion years ago. Second, magnetotaxis requires the presence of a strong planetary magnetic field. On Earth, a strong early field probably decayed after 3.5 billion years ago, only to reach its present level after nucleation of the inner core about 2.8 billion years ago.

Kirschvink has thus postulated a novel scenario—and perhaps the most plausible scenario—for the formation of the eukaryotic cell: a pathway necessitating the presence of magnetite and a strong planetary magnetic field. As we shall see in a later chapter, not all planets maintain magnetic fields. If this pathway is the *only* way to large eukaryotic cells (a hypothesis that still awaits verification), then we have another requirement we must impose on planets that aspire to host animal life—a magnetic field.

Environmental Conditions Leading to the Evolution of Eukaryotes

What environmental conditions led to the evolution of the forerunners of animal life? New discoveries of the 1980s and 1990s have given us a much clearer view of the early Earth during the great evolutionary transitions we saw in the last chapter. The Earth's earliest life seems to have formed during or soon after cessation of the heavy comet bombardment. By about 3.8 billion years ago that heavy cosmic bombardment ended, and by 3.5 billion years ago we find the first fossilized evidence of life.

The region that has yielded Earth's oldest fossils found to date is known to Australians as the North Pole, because even in this isolated continent, it is uniquely remote and inhospitable. The rocks in this region belong to a unit of interbedded sedimentary and volcanic rocks known as the Warrawoona Series. Geologists have deduced that the deposits were consolidated in a shallow sea over 3.5 billion years ago. There is the evidence of storm layers and evidence as well that on occasion, a hot sun evaporated small pools of seawater into brine deposits. But it is not these structures that have created so much excitement about the Warrawoona rocks. This ancient bit of Australia holds the world's oldest stromatolites, low mounds of lime and laminated sediment that have been interpreted as the remains of microbial mats—in other words, life.

Stromatolites (the "stone mattresses" we mentioned earlier as an anomaly, multicellular prokaryotes) are the most conspicuous fossils and the most commonly preserved evidence of life for more than 3 billion years of Earth history; they provide our best record of early life. They have been found on every continent in rocks half a billion years old and older. Today, they are found in only one type of environment on Earth, in quiet, briny tropical waters. Such environments are refuges from algal grazers; stromatolites can no longer exist on most of our planet's surface because they would quickly be eaten. The photosynthesizing bacteria termed cyanobacteria are modern equivalents of these ancient deposits.

The presence of stromatolites is a sure clue that by 3.5 billion years ago, life on this planet had left its earliest, probably hydrothermal or deep-earth environments and diversified onto the surface of the planet. For a billion years the prokaryotes were masters of the world, but life was still scattered. According to the fossil record, it was not until about 2.5 billion years ago that the organisms that produced stromatolites had released sufficient quantities of oxygen to form sedimentary deposits known as banded-iron formations. Prior to the appearance of common stromatolites, there was no dissolved oxygen in the sea, no gaseous oxygen in the atmosphere, and hence no possibility of mineral oxidation. With the appearance of oxygen, however, large volumes of iron that had been dissolved in seawater precipitated out as it oxidized into iron oxides—rust, in other words. Today, there still exist at least 600 trillion tons of such iron oxides deposited before 2.5 billion years ago in these banded-iron formations.

The time interval commencing about 2.5 billion years ago is marked by a change in the very tectonic nature of the planet Earth—its rate of mountain building and continental drift. By this time, the heat production from radioactive elements locked in Earth's rocks had diminished, for some of the radioactive elements decayed rapidly early in Earth's history. This material was like a finite amount of fuel within the interior of the planet, and as it was used up, heat flow declined. It turns out that the processes of continental drift and mountain building are by-products of heat rising from within Earth, and as the amount of heat decreased over time, so did these two activities. There is also some evidence that around this time, a major pulse of land formation occurred, allowing larger continental land masses to form. As the new continents formed, many shallow-water habitats were created, and these proved favorable environments for the growth of photosynthesizing bacteria. We can speculate that from about 4 billion to about 2.5 billion years ago, there were few large continents, but numerous volcanic island chains dotted the world. After 2.5 billion years ago, continental land masses began to form, and volcanism on a global scale lessened.

With this increase of habitat, ever more stromatolites grew and flourished. This in turn relentlessly pumped ever more oxygen into the sea. As long as there was dissolved iron in the seawater, all of the liberated oxygen was quickly locked up in the banded-iron formations. By about 1.8 billion years ago, however, this reservoir of dissolved iron material was used up. We know this because after that time, no more banded-iron formations were laid down. This changeover left an indelible mark on the sedimentary record of Earth, for as the sea became saturated with oxygen, the time of banded-iron formations ended forever—or at least until some far-distant future when our planet may again no longer have oxygen. With nowhere else to go, oxygen began to emerge into our planet's atmosphere and, in so doing, probably gave life its first impetus toward *animal* life.

THE OXYGEN REVOLUTION

It is probably impossible for us to conceive how entirely alien to ours this world truly was. Yet the strange microbial world of 2 billion years ago may be the norm in the Universe for those planets that harbor life. Traces of it exist still, here on Earth, in the bacterial froths and pond scum that persist across our planet, and perhaps nowhere more prolifically than in the rotting garbage dumps and landfills created by our own species—places where huge, visible colonies of rapidly growing bacteria exist still. But the rainbow slick of the oozing swamp is the exception in a world where the eucaryans are so much more in evidence than the prokaryotic forms. What would that 2-billion-year-old world look like? The best description we know of was penned by two scientists who have journeyed back to this world, in their imaginations, many times. We owe the following image of the ancient Proterozoic era (the formal name for the time interval of 2.5 to 0.5 billion years ago) to Lynn Margulis and Dorion Sagan, in their 1986 book *Microcosmos*:

> To a casual observer, the early Proterozoic world would have looked largely flat and damp, an alien yet familiar landscape, with volcanoes smoking in the background and shallow, brilliantly colored pools abounding and mysterious greenish and brownish patches of scum floating on the waters, stuck to the banks of rivers, tainting the damp soils like fine molds. A ruddy sheen would coat the stench-filled waters. Shrunk to microscopic perspective, a fantastic landscape of bobbing purple, aquamarine, red, and yellow spheres would come into view. Inside the violet spheres of *Thiocapsa*, suspended yellow globules of sulfur would emit bubbles of skunky gas. Colonies of ensheathed viscous organisms would stretch to the horizon. One end stuck to rocks, the other ends of

some bacteria would insinuate themselves inside tiny cracks and begin to penetrate the rock itself. Long skinny filaments would leave the pack of their brethren, gliding by slowly, searching for a better place in the sun. Squiggling bacterial whips shaped like corkscrews or fusilli pasta would dart by. Multicellular filaments and tacky, textilelike crowds of bacterial cells would wave with the currents, coating pebbles with brilliant shades of red, pink, yellow and green. Showers of spheres, blown by breezes, would splash and crash against the vast frontier of low-lying mud and waters.

This prokaryotic world was creating what has been called the Oxygen Revolution. The initiation of an oxygenated atmosphere was one of the most significant of all biologically mediated events on Earth. Prokaryotic bacteria, using only sunlight, water, and carbon dioxide, ultimately transformed the planet by generating an ever-increasing volume of atmospheric oxygen. This outpouring of oxygen created both biotic opportunity and biotic crises. Many of Earth's primitive organisms were metabolically incapable of dealing with abundant oxygen. For most of the archaeans, the oxygen boom of about 2 billion years ago was an environmental disaster, driving some species into airless habitats, such as lake and stagnant ocean bottoms, sediments, and dead organisms. Others were incapable of such migration and simply died out. For yet other creatures, however, the profound change in atmospheric conditions created new opportunities. Some prokaryotic cells began to exploit the enormous power of oxygen metabolism to break down food sources into carbon dioxide and water. This new metabolic pathway yielded far more energy than any of the anaerobic pathways. Organisms that adopted it soon began to take over the world. The most efficient of these were members of the domain Eucarya, which, more than 2 billion years ago, evolved true eukaryotic cell machinery.

The oldest known fossils of an organism that appears to have attained the eukaryotic grade of organization have been found in banded-iron deposits located in Michigan. The fossils themselves are about 1 millimeter in diameter and are found in chains as much as 90 millimeters long. The organism, then, is far too large to be a single-celled prokaryote or even a single-celled eukaryote. This creature, which has been named *Grypania*, is preserved as coiled films of carbon on smooth sedimentary rock bedding planes, the places where sedimentary beds split apart. Its 1992 discovery indicates that the evolution of the first eukaryotic cell occurred *during* the banded-iron formation process, when there was still little free oxygen in the sea and probably none in the atmosphere. These early eukaryotes may have been vanishingly rare, for other eukaryotes do not occur in the fossil record for 500 million years after this first appearance, but with this form, a beachhead in life's advance had been established.

For the period between 2 and 1 billion years ago (see Figure 5.2), few notable achievements of life are recorded as fossils in the rocks. The first common appearance of eukaryotes begins about 1.6 billion years ago, when microscopic fossils called acritarchs begin to appear in the geological record. These are spherical fossils with relatively thick, organic cell walls. They are interpreted to be the remains of planktonic algae, forms that used photosynthesis and lived in the shallow waters of the world's oceans. Other life forms evolved as well, but as is also true of most living protists, such as the amoeba and the paramecium, their lack of skeletons renders them invisible in the fossil record. With a proliferation of plant-like forms, new varieties of predatory protists surely evolved. Whole armadas of single-celled, floating pastures and the somewhat larger and more mobile grazers on these fields of plankton lived and died in this seemingly endless epoch of geological time. The open ocean would have had little life, but the coastal regions richer in nutrients would have been awash with life—microscopic life. It was the Age of Protists, the Age of the Small.

We have now reached 1 billion years ago, in our march through evolutionary time. Finally, the tempo of evolutionary development increased, if we are correctly interpreting the fossil record, for there is a burgeoning in the number of eukaryotic species found in the rock record at this time. Some of these new forms include the first red and green algae, forms still crucial and varied in marine ecosystems. This diversification of eukaryotic species, including protozoans and plants, set the stage for the evolution of larger, multicellular forms and may have been triggered by the evolution of important new morphologies within the eukaryotic cell.

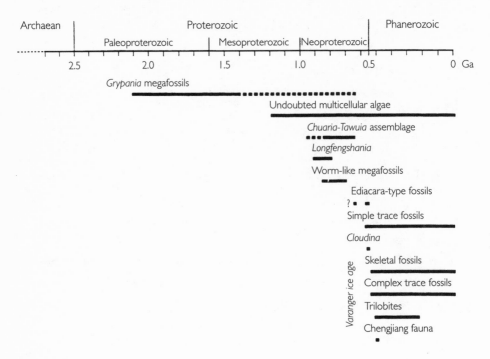

Figure 5.2 *Early multicellular fossils. Broken bars indicate uncertain time ranges.*

EVOLUTION IN EUKARYOTIC FORM AND FUNCTION

Four biological innovations may have been especially significant in paving the way for the emergence of larger animals: (1) the development of sexual cycles; (2) new methods of shuffling information coded along the chromosomes (through the nascent ability to excise and relocate entire gene sequences); (3) new methods of communicating between cells via substances called protein kinases, and (4) the development of a new type of intracellular skeleton, called a cytoskeleton, that allowed eukaryotic cells to increase enormously in size. These innovations greatly enhanced the ability of cells to

100

evolve new morphologies in response to natural selection and their ability to band together into multicellular creatures.

We can now better categorize what we term "advanced" life: eukaryotic multicellular organisms. There are, of course, many types of *multicellular* organisms, including a considerable number of prokaryotic forms. In most cases these multicellular prokaryotes are composed of only two cell types. Cellular slime molds are multicellular, as are some cyanobacteria. In a way, however, these forms are evolutionary dead ends. They have existed on Earth for several billion years and are highly conservative in an evolutionary sense. It is multicellular creatures of the other category that became so important in the history of life. We refer here to true metazoans.

The jump from single-celled organisms to organisms of multiple cells requires numerous evolutionary steps. The jump from single-celled organisms to metazoan *animals*, where a high degree of intercellular cooperation in organization exists, involves even more. In their recent book *Cells, Embryos and Evolution*, biologists John Gerhart and Marc Kirschner discuss this evolutionary accomplishment. The first step, they argue, seems almost paradoxical: It was not some new structure gained that allowed this transition, but an important structure lost. Long ago in our planet's past, some organism of the eukaryotic lineage made a brave (or lucky) morphological change—it shed its external cell wall. Why this occurred is still unclear, but the net effect was far-reaching. A tough outer coating protects most unicellular creatures from their surrounding environment. At the same time, however, it isolates these cells from other members of their own kind. By divesting themselves of this outer wall, individual cells could begin exchanging living material—and information—with one another. The naked cells could adhere to each other, crawl over each other, and communicate. These were the first steps in the formation of a tissue, which is an aggregation of cells united for mutual benefit.

Larger animals require highly integrated systems of cells that can accomplish the myriad functions necessary for all life. Respiration, feeding, reproduction, the elimination of waste material, information reception, locomotion—all require the integration of many cells acting in concert. Each of these functions ultimately requires one or many types of tissues.

Among tissue types, the outer wall of any organism (the epithelium) is of utmost importance. The epithelium must protect the organism from the rigors of the external environment but at the same time allow adsorption of critical gas and, sometimes, nutrients. The evolution of the epithelium was a decisive first step in the evolution of metazoans.

Which group of unicellular creatures first achieved this breakthrough? The most primitive and enigmatic of larger eukaryotic metazoans are sponges. These curious creatures seem to bridge the gap between single-celled eukaryotes or even colonial protozoa and the highly integrated invertebrate metazoan phyla. Sponges have several cell types that perform specialized tasks, but there is a very low level of organism-wide organization. There is no gut or body cavity specialized for processing food, nor is there any nervous system. Yet the sponges may be an important clue to the identity of our actual metazoan ancestors.

The stem, or ancestral, metazoan probably had a larger number of cell types than sponges (perhaps 10 to 15 rather than the 3 to 5 individual cell types found in sponges). There was probably a body cavity of some sort segregated into two cell layers: an outer ectoderm and an inward-facing endoderm. This two-tissue plan seems to have been an evolutionary dead end, and it wasn't until a third layer—the mesoderm—was added that animals with real internal complexity formed. Eventually, a small worm-like shape with three tissue layers evolved, a creature with a gut running through the long axis of the body and a separate space known as a coelom to serve as an internal hydrostatic skeleton. With this tiny organism (the first may have been less than a millimeter long), the evolutionary stage was set for the emergence of animals on planet Earth.

THE TWO DIVERSIFICATIONS
OF ANIMAL PHYLA

With the advent of this form—what evolutionary biologists call the "roundish flatworm"—a body plan was in place that could be modified to shape all the categories of metazoan life, the major body plans that we call phyla. The phyla living today include the arthropods; the mollusks; the echinoderms; our own group, the chordates; and about 25 more. These are the complex metazoans that we hope to find—but that may be very rare—on other planets. These are animals. They appeared relatively late in the history of life on Earth. One of the great novel insights of the 1990s was our realizing that their origin and their subsequent diversification and rise to abundance were two separate events, not one, as had been believed since the time of Charles Darwin.

Fossils of macroscopic animals (those visible to the unaided eye) first appear in abundance less than 600 million years ago, during the "Cambrian Explosion," a diversification event resulting in the rapid formation of thousands of new species; we will describe it in more detail in the next chapter. Yet the appearance of abundant animal fossils at this time actually marks the second of the two diversification events that led to the proliferation of larger animals on the planet. As we will show, fossils of such complex animals as trilobites and mollusks—common members of the Cambrian Explosion—are advanced descendants of a much earlier, diversification event that took place between 1 billion and 600 million years ago. Yet there is no *fossil* record of this first diversification—paleontologists have been stymied by an almost compete lack of fossils in strata older than 600 million years, when this initial event must have taken place. Our understanding of the initial diversification of animals comes not from paleontology but from an entirely different line of investigation: genetics. Geneticists have arrived at answers about the "when" of the first diversification event by examining the genetic code of living animals via a technique called ribosomal RNA analysis.

Gene sequences are simply strings of base pairs lined up along the double helix of a DNA molecule. As we saw earlier, if a DNA molecule is likened to a twisted ladder, the base pairs can be considered the steps of the ladder, and it is the sequence of the steps that is used in this type of analysis. Genes are simply instructions for protein formation coded by the sequence of nucleotides on the DNA ladder. There are only four types of nucleotides, but they provide the genetic code that is the basis for all Earth life. All organisms share more genes with their ancestors than with nonrelated species. By comparing the genes from various organisms, it is possible to produce a model of evolutionary history (an evolutionary tree, as it were) with the branches of the tree showing which species gave rise to which other species. Yet according to many geneticists, such an analysis not only tells us how the branching occurred, it can also tell us when.

In 1996 G. Wray, J. Levinton, and L. Shapiro published a paper claiming, on the basis of results obtained by using this genetic technique, that the first event—the earliest divergence of animals—occurred 1.2 billion years ago. This result drew a collective gasp from the paleontological fraternity: It seemed much too ancient. The fundamental assumption of the Wray *et al.* paper is that gene sequences evolve with sufficient regularity that a sort of molecular "clock" can be used to date the divergence of various groups. The reasoning behind the molecular clock technique is that changes in the genetic code—evolution, in other words—occur at a rather constant rate. The more distinct two DNA sequences are, the longer it has been since they diverged from a common ancestor. Other scientists, however, dispute that changes in gene frequency occur at a constant rate, and therefore they do not believe in the molecular clock. It is these molecular clock data that led the Wray group to their conclusion. This finding was a bombshell. If animals evolved this early, why did they not appear in the fossil record until less than 600 million years ago? What were they doing for such a long time?

The Wray group's findings were extremely controversial not only because they contradicted long-held paleontological dogma but also because they provoked criticism among other geneticists. There is fierce debate among geneticists about the reliability of the molecular clock technique. The Wray study itself, yielded both minimum and maximum figures for the earliest divergence. One group of genes suggested that the fundamental splitting of the phylum made up of annelids (worms) from the phylum of chordates (our phylum) occurred only 773 million years ago, whereas a second group of genes (in the same organisms) suggested 1621 million years ago—a very wide spread indeed! These results give us minimum and maximum ages for the divergence. Even with the minimum figure, however, there were (according to the molecular data, anyway) recognizable chordates and annelids 700 million years ago—yet there is no trace of their presence in the fossil record. Where were they? Or were they not there at all? Could it be that no rocks of this age survive or that no fossils from the interval of about 1 billion to less than 600 million years ago were preserved? This seems to be stretching things, as suggested by British paleontologist Simon Conway Morris:

> Appeals to gaps in the rock record and pervasive metamorphism of the sediments are not going to work: if there were large metazoans capable of either fossilization or leaving traces, they had an uncanny knack of avoiding areas of high preservation potential.

Since the original, tantalizing analysis by the Wray group, other geneticists have reconsidered the basic data. Most concede that the 1.2-billion-year figure is too old. (However, a report published in *Science* magazine in late 1998 by a team headed by Dolf Seilacher of Yale University announced the discovery of billion-year-old trace fossils (worm-tracks) possibly derived from small, worm-like organisms. Critics of this finding suggest that the marks in question could just as easily have been produced by inorganic actions, and even if these trace fossils turn out to have been produced by organisms, the question remains: Why are no further such fossils found for hundreds of millions of years?) Let's say, then, that divergence occurred less than a billion years ago. We must still account for a significant period of time with animals but without fossils. Paleontologists have long believed that only a single major diversification event occurred—the event coincident with the appearance of fossils, the so-called Cambrian Explosion that began about 550 million years ago. Now this evolutionary event is seen as a follow-up to the much earlier first event.

The answer to this seeming conundrum is that the animals were indeed present, but they were so small as to be essentially invisible in the fossil record. A recent and spectacular discovery of microscopic fossil animal embryos seems to confirm this view. Using newly developed techniques of searching for tiny (but complex) animals in minerals called phosphates, paleontologist

Andy Knoll and his colleagues have uncovered a suite of tiny but beautifully preserved fossils interpreted to be the embryos of 570-million-year-old triploblasts—animals with three body layers, like most of those found today. These fossils tell us that the ancestors of the modern phyla were indeed present at least 50 million years before we find any conventional fossil record of them. The combination of genetic information and new discoveries from the fossil record now give us a robust view of the rise of animals: They did not exist 1 billion years ago, and perhaps not 750 million years ago. Animals are indeed very late arrivals on the stage of life on Earth.

Thanks to these new discoveries and interpretations, the question of "when" has been answered to most people's satisfaction: The emergence of animals was a two-stage event. The initial stage seems to have occurred less (and perhaps much less) than the billion years ago proposed by Wray and his colleagues. But even recalibrated, the Wray group's finding has given us yet another tantalizing insight into the potential incidence of animal life in the Universe. The Wray work confirms that there were indeed two "explosions." The first was the actual differentiation of the various body plans; the second was the differentiation and evolution, in these various phyla, of species large and abundant enough to enter the fossil record. The geneticists can show that genes of annelid worms and genes of chordates were differentiating hundreds of millions of years before the emergence of these creatures as large entities that could appear in the fossil record. This leads us to ask a crucial question: Even if they evolve, *do animals necessarily, or inherently, go on to diversify, enlarge, and survive?* Does the second flowering of animal life—the Cambrian Explosion event so long known to geologists—inevitably follow the first diversification, or is it yet another threshold of possibility that may be (but is not necessarily) attained? Perhaps on some worlds in the Universe, animals diversify but never attain larger size and greater numbers in some Cambrian Explosion equivalent. This particular insight was first expressed by paleontologist Simon Conway Morris:

> We need to discuss to what extent metazoan history was *implicit* a billion years ago, at least in outline, as opposed to what was inevitable 500 million years later at the onset of the Cambrian ex-

106

plosion. Even if metazoans have a deep history, which paleonto-logically remains cryptic, the actual organisms would have been of millimeter size and perhaps without the potential for macroscopic size and complex ecology. . . . Wray *et al.* may have been correct in tracing the gunpowder back as far into the mists of the Neo-proterozoic (the late Precambrian time period of a billion years ago), but the keg itself still looks as if it blew up in the Cambrian.

In other words, it seems that the development of animals was a two-step process, with step two—the Cambrian Explosion—not necessarily being an outcome predetermined by the initial differentiation of the animal phyla.

Over and over the same question arises: Why did it take so long for animals to emerge on planet Earth? Was it due to external environmental factors, such as the lack of oxygen for so long in the history of this planet, or to biological factors, such as the absence of key morphological or physiological innovations?

THE EVOLUTION OF ANIMALS: BIOLOGICAL BREAKTHROUGH OR ENVIRONMENTAL STIMULUS?

Complex animals surely cannot appear on any planet without following some evolutionary pathway from simpler, single-celled organisms. The change from single-celled microbes to multicellular creatures must be the common route on any planet, and even if the molecules of life are different from world to world, the pathway from simple to complex may be universal. Because of this, the example of how animals evolved on our Earth may be of the utmost importance in understanding the frequency with which animals occur on other planets.

If we are to understand how animals evolved from single-celled ancestors, we must first understand the environments where these monumental evolutionary advances were made. We know well the "when" of this change—it took place during a 500-million-year interval from 1 billion to 550 million years ago. The second event, the Cambrian Explosion of between 550 and 500 million years ago, included the morphological diversification of the phyla into subdivisions based on body plans, as well as the appearance, within the various phyla, of species with skeletons and large size (see Figure 5.3).

During this interval of time, Earth went though major environmental changes, among them ice ages of unprecedented severity, rapid continental movements, and drastic changes in ocean chemistry. We are thus left with perplexing questions: Did the environmental changes of this interval (which are described in more detail below) somehow trigger the diversification of an-

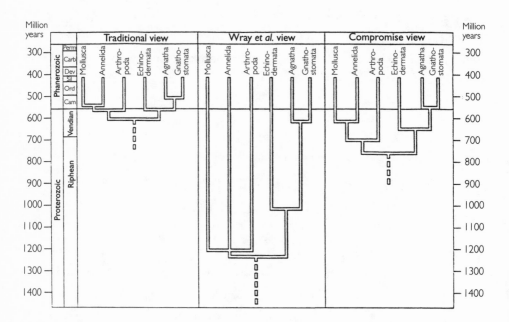

Figure 5.3 Differing views of metazoan phylogeny. Most paleontologists follow the "traditional view" (left), accepting the fossil record as a fairly reliable indicator of original events. Molecular clocks are interpreted by Wray et al. (center) as indicating very deep origins for the principal metazoan phyla. The recognition that some molecular clocks run much faster than others suggests a "compromise view" (right), which implies that our search strategy for the first metazoans should be concentrated in the interval from about 750 million years onward. Perm, Permian; Carb, Carboniferous; Dev, Devonian; Sil, Silurian; Ord, Ordovician; Cam, Cambrian.

imals? Or would the rise of animals have occurred even in the absence of these profound environmental changes? These questions, which of course are central to understanding the evolution of life on our planet, have great relevance to understanding the frequency of animal life on other planets as well. Does animal life *always* (or even commonly) evolve, once a suitable ancestor appears? Or does there need to be an additional trigger of some sort, a sequence of environmental steps? We might compare this whole process to baking a cake. Say the ingredients for the batter were all assembled and mixed by 1 billion years ago. Does the cake need to be cooked for a given time at a highly restricted temperature in order to rise? Or will any amount of cooking at any temperature accomplish the task just as well? Or will our cake be completed without any cooking at all? (That is, does simply assembling the ingredients into a batter ensure success?)

The beginning of this fecund period in Earth history is marked by the appearance not of new types of animals, but of plants. Around 1 billion years ago, many types of algae begin to appear in the fossil record, including the green and red algae still so prominent on Earth today. These were not the ancestors of animals, of course, but their appearance was the opening salvo of an evolutionary assault that was the most significant up to that time. It was followed, hundreds of millions of years later, first by the initial diversification of animal phyla and then (after more hundreds of millions of years) by the Cambrian Explosion of animal life.

What were the environmental events of this interval of time 1 billion to 600 million years ago? By this period, land masses approaching the size of today's continents had formed, and the total area of land on the planet may not have been significantly different from what we see in the present day. The land, however, was not a tranquil place. The period was one of significant mountain building and continental drift. It was also marked by episodes of continental glaciation unmatched in severity since that time. Did these events have anything to do with the diversification of animals? One school of thought says yes. Work by Martin Brasier and others suggests that rapid changes in sea level, and especially the formation of broad, shallow seas within the new continents, would have opened up many new habitats very hospitable in terms of temperature and nutrients. This, in turn, may have stimulated the diversification of animals and plants. There are dissenters, notably James Valentine, who cautions, "the link between plate tectonics . . . and the origins and radiation of animals remains to be demonstrated." But, as Harvard paleobiologist Andy Knoll points out, there is another way in which the new and active tectonic events could have influenced the initial radiation of animals that occurred during this time. In 1994 Knoll noted that "tectonic processes could have influenced one or more of the great radiations (of animals) . . . through their participation in the biogeochemical cycles that regulate Earth's surface environments."

Examples of such effects include the role of hydrothermal influences on ocean chemistry. The hydrothermal vents, as we saw in Chapter 1, are submarine regions where great volumes of hot and chemically distinctive water are mixed with seawater. The amount of this volcanically derived water entering the oceans fluctuated during the interval of 1 billion to 550 million years ago, and these fluctuations had marked effects on the chemistry of the seawater, on the composition of the atmosphere, and on climate. The tectonics events also affected the rate of burial and exhumation of organic carbon in sediments. Oxygen and carbon dioxide values shifted, and as they did so, major changes in the temperature and oxygenation of the planet ensued.

Yet another environmental stimulus may also have contributed to the initial animal diversification. Changes in ocean chemistry caused by increased tectonic activity beginning a billion years ago facilitated the evolution of skeletons. This period is marked by the appearance of rocks called phosphorites. Some authors credit these rocks with bringing about an increase in the fertility of the oceans at this time, which may in turn have helped trigger the sudden appearance of many diverse animals beginning about 600 million years ago. Phosphorus is much more concentrated in living things than in the environment, so it is a limiting nutrient. The sudden presence of abundant sources of this element could have acted as a veritable fertilizer for growth.

Knoll has discussed all of these disparate factors and has proposed three alternatives. First, it may be that the complex physical events and the equally complex series of biological events that occurred from 1 billion to 550 million years ago are simply coincidental—they had nothing to do with one another. If this is true, the great biological diversification must be attributed solely to biological innovations (such as the ability of cells to bind together, build an outer cell wall, and evolve internal cooperation between contained cells) that were unrelated to concurrent environment changes.

The second alternative is that evolution was indeed facilitated by changes in the physical environment. The most important of these changes may have been in levels of oxygen. The first appearance of larger metazoans, the ediacarans, about 600 million years ago occurred immediately after a sudden increase in atmospheric oxygen (evidence for this comes from stable isotopes). Thus it may be that the initial animal diversification of around 700 million years ago was itself a response to the oxygen level reaching some critical threshold.

The third alternative is that the biological revolutions themselves somehow triggered some of the physical events—just the opposite of alternative two! In this scenario, the common use of calcium carbonate shells by newly evolved animals changed the way calcium was distributed in the oceans. Similarly, organisms may have favored the formation of phosphorus, not the other way around: The presence of many organisms may have changed the physical chemistry of the ocean environment, boosting the formation of this mineral type.

Knoll leans toward the last alternative. He stresses that the first major evolutionary radiation among protists and algae (about 1 billion years ago) may have occurred because of the first evolution of sexual reproduction. The invention of sex, rather than an environmental trigger, stoked the fires of diversification. But Knoll also acknowledges the central role of oxygenation in the evolution of larger animals. Without oxygen, larger animals could never have evolved, and oxygenation during this interval was facilitated by tectonic processes—specifically, the role of changes in sea level and erosion of continents in complex geochemical cycles. For a variety of physiological reasons, oxygen is a key to the appearance of larger animals; the metabolism of animals requires oxygen.

Indeed, we may well ask whether oxygenation, and hence the rise of animals, would *ever* have occurred on a world where there were no continents to erode. Perhaps "water worlds" are ultimately inimical to animal life. But there may have been even more sudden and catastrophic changes than those listed by Knoll—most important, dramatic changes in planetary temperature. Evidence uncovered in the late 1990s has led to a radical new concept: that Earth almost completely froze over at least twice in its history—once 2.5 billion years ago and a second time (perhaps repeatedly) during the interval from about 800 to some 600 million years ago. These times of intense global cold, when even the oceans were covered with ice, are known as Snowball Earth. Their biological significance is explored in the next chapter.

Snowball
Earth

"Let it snow, let it snow, let it snow."

—Christmas song

It is hard to hide our genes completely.

—Philip Kitchner, *The Lives to Come, 1996*

Spring is universally associated with birth, growth, and fertility. It is a time of warmth and renewal after the frigid lifelessness of winter. And so it would seem that the emergence of animals long ago on Earth should have resulted from a protracted period of warm and fertile, spring-like conditions. But new information uncovered by several insightful scientists suggests that the birth of animal life on Earth was initiated not by a time of warmth but, rather, by the most fearful winter ever to grip the planet. If this phenomenon, known as Snowball Earth, turns out to be linked to the origins of animal life, what will it mean for the possibility of animal life on other planets?

As we noted earlier, a majority of astrobiologists believe that the temperature of early Earth from the time of the first life, about 3.8 billion years ago, until the origin of eukaryotic cells, about 2.5 billion years ago, was high—probably too hot for the existence of animal life. (Yet there are others who suggest that Earth may have undergone a "cool start," because the sun at that time was giving off much less energy than now). Both camps agree that the planet's atmosphere was almost devoid of oxygen. Those who believe in a "hot start" suggest that gradually, as greenhouse gases in the atmosphere were reduced in volume, the temperatures declined. But Earth may have cooled too much (or, if you are of the "cool start" persuasion, failed to warm up enough), at least in the short term. There is evidence of as many as four major episodes of glaciation on a scale far exceeding anything before or since—times of cold and ice that make the last ice age, the Pleistocene epoch of 2.5 million to 10,000 years ago, seem but a brief cold snap.

The first known Snowball Earth episode began about 2.45 billion years ago, and a second protracted siege of several such events occurred between 800 and 600 million years ago. These two dates are of great interest, because they are also the times of the two most signal events in biological history since life's first appearance here: Around 2.5 billion years ago the first eukaryotic cells appeared, and the fossil record reveals that about 550 million years ago, diverse and abundant animal life blossomed, in the event known as the Cambrian Explosion, the subject of the next chapter. Perhaps it is just coincidental that these two spectacular and far-reaching biological events occurred immediately after the two most severe episodes of glaciation and ice cover in Earth history. But according to a controversial new theory, both may have been triggered by the Snowball Earth episodes.

IMPRISONED IN ICE

Continental glaciations leave evidence of their former presence: a characteristic topography on the landscape, grooves and scratches caused as the passing glaciers ground over hard rock, and (perhaps most important) telltale sedimentary deposits called tillites. The latter are deposits of angular rock fragments, which were carried and then left by moving glaciers. The recently concluded ice ages of 2.5 million to 12,000 years ago left many such deposits in both the Northern and the Southern hemispheres. Such tillite deposits are also found in much older rocks. Thick tillite deposits have been recovered from two different intervals in Precambrian Earth history: around 2.4 billion years ago and during the interval from about 800 to 650 million years ago. The unusual aspect of these features is that they are recovered from virtually all latitudinal regions of the globe, which shows that the glaciations extended to near equatorial latitudes (in contrast to the more recent glaciations, which extended from the poles to mid-latitudes). It may be that no region then on Earth escaped the glaciation. So much of the planet was covered by ice in these two Precambrian ice ages that in 1992, Dr. Joseph Kirschvink of Cal Tech dubbed them "Snowball Earth" events. Far different from the later ice ages, they were times when Earth teetered dangerously close to becoming too cold for any life. The Snowball Earth theory received a boost in August 1998 with Harvard geologist Paul Hoffman's publication, in *Science,* of new evidence that ice extended to near equatorial latitudes in the late Precambrian, about 700 million years ago.

The more recent glaciations, those that occurred since skeletons evolved about 550 million years ago, affected only land regions; except for an increase in icebergs, or at most ice cover near the continents, the oceans remained open. Such may not have been the case in the Precambrian glaciations. During these two "Snowball Earth" episodes, all of the oceans may have been covered with ice to considerable depths. And although the deeper regions of the seas remained liquid, thick icebergs, or pack ice to depths of 500 to 1500 meters, may have covered the ocean. Earth would have been cold indeed. Average surface temperatures on the planet would have varied between $-20°C$ and $-50°C$.

These extremely cold temperatures would have had an enormous influence on the surface of our planet. For example, continental weathering would have slowed or even stopped. In the interior of continents, the covering of ice would eventually ablate (evaporate) away, just as it does in the dry valleys of Antarctica today, leaving behind a sterile rock surface. Dust from these regions would be blown out to sea, making the pack-ice cover of the oceans brown from terrigenous material. From space, Earth would have looked white and brown—the white being the ice covers on the oceans, the brown the denuded land areas.

The presence of the pack ice covering the sea would act as a lid on a pot. Normally, much free exchange occurs across the vast interface of ocean and atmosphere. Water evaporates from the sea into the air and then rains back into the sea. If the sea were covered with ice, however, the ocean and the atmosphere would become "decoupled." Chemical changes in the ocean would be separated from the atmosphere by the kilometer-thick lid of ice on the ocean surface. Very drastic chemical change could—and according to Kirschvink and others, did—occur within the sea itself.

Even with the icy cover, volcanism would have continued both on the land's surface and along the mid-ocean volcanic ridges at the bottoms of the world's oceans. At such sites today (see Chapter 1), great volumes of metal-rich fluids gush forth from these submarine volcanoes. In a covered ocean, this material would have become toxic, producing what are known as reducing conditions. The oceans would have begun to accumulate with metal ions, mainly iron and manganese. For as long as 30 million years, the glaciers and ice never relaxed their frigid grip on the planet's surface.

All of this global cold would surely have adversely affected life in the shallow-water regions of all the world's oceans. The biosphere became restricted to a narrow belt around the equator and to deep-sea hot springs and hydrothermal vent settings. Perhaps some life also survived in occasional Yellowstone-like hydrothermal systems.

Astronomers once thought that a previously warm world's descent into such an "icehouse" or "snowball" would be irreversible. Their reasoning was that as a planet gets more and more thickly coated by ice, the fraction of sunlight reflected back into space increases and solar heating of the surface declines. On Earth today, sunlight is adsorbed by the darker land and seas but is reflected into space by cloud cover. A planet completely covered with ice would reflect *most* sunlight into space, causing the planet to become ever cooler. Yet it is clear that Earth was able to escape from the deep freeze—not once but several times. The means of that escape was through the volcanic emissions of greenhouse gases such as carbon dioxide into the atmosphere, producing a "greenhouse effect."

ESCAPE

As we saw in Chapter 2, a planet's average temperature is greatly affected by the volumes of greenhouse gases in its atmosphere. Much of this gas enters a planet's atmosphere from actively erupting volcanoes. Although there are abundant volcanic eruptions in the sea as well, most of the carbon dioxide from these events does not make its way into the atmosphere. Cold seawater can hold large amounts of dissolved carbon dioxide, and below 700 meters, CO_2 will settle to the bottom of the ocean as it reaches saturation in the water. At the time of Snowball Earth, enough CO_2 would eventually reach the atmosphere to melt back the sea ice and, in so doing, expose the metal-rich waters of the sea to the atmosphere. The time necessary for this "melt-back" has been estimated by Hoffman and his group to be between 4 and 30 million years. With the ice melted back from the sea, and temperatures again warming, Earth would have undergone spectacular changes. Here is how Kirschvink has described these events:

> Escape from this "icehouse" condition was only accomplished by the buildup of volcanic gases, particularly carbon dioxide, mostly from undersea volcanic activity. Deglaciation during the end of these glacial events must have been spectacular, with nearly 30 million years of carbon dioxide, ferrous iron, and long buried nutrients suddenly being exposed to fresh air and sunlight. Hundreds of meters of carbonate rock are preserved capping the glacial sediments, at all latitudes, on all continents, as a direct result of wild photosynthetic activity. For a brief time, the Earth's oceans would have been as green as Irish clover, and the sudden oxygen spikes may have sparked early animal evolution.

The most important source of biological productivity in the oceans of today derives from the growth of phytoplankton, the single-celled plants that are the pastures of the sea. The growth of these plants, so important for producing oxygen, is limited by the availability of nutrients and iron. If iron is dropped into the oceans of today, a great bloom of phytoplankton results. Such was probably the case soon after the end of the first Snowball Earth event. As the ice-covered seas began to melt, the fine iron- and magnesium-rich dust coating the surface of the sea ice would have acted as a fertilizer, tremendously stimulating growth of the blue-green "algae" (really photosynthesizing bacteria known as cyanobacteria). Enormous populations of cyanobacteria would have clotted the surface regions of the liberated seas, releasing huge volumes of oxygen as a consequence of their photosynthetic activity. This sudden appearance of so much life, after the millions of years of cold and dearth of life, would have been a great revolution, and it probably stimulated new evolutionary changes.

These events would have had profound geological as well as biological ramifications. The sudden rush of oxygen into the sea and air would have caused the iron- and manganese-rich oceans to precipitate out iron and manganese oxides. In a previous chapter we saw how banded-iron deposits began to accumulate about 2.5 billion years ago. Kirschvink and his group argue that the appearance of banded-iron deposition occurred soon after the first Snowball Earth ended. Not only iron deposits but magnesium-rich deposits as well were immediate results of the end of the first Snowball Earth event. Evidence of this is seen in South Africa, where the world's largest land-based deposit of manganese minerals has been dated at 2.4 billion years of age and sits just above sedimentary deposits that were laid down during the 2.5-billion-year-old Snowball Earth episode. Like the banded-iron formations, these manganese-rich deposits appear to be a direct consequence of the oxygen bloom that occurred when the planetary snowball melted.

The cessation of the 2.5-billion-year-old Snowball Earth thus appears to have resulted in a rise in the amount of oxygen both dissolved in the sea and free in the atmosphere. Probably for the first time in Earth's history, the sunlit portions of the sea became too oxygen-rich to allow iron to exist in solution in seawater. Kirschvink and his colleagues argue that this dramatic change in the chemistry of the sea would have exerted intense evolutionary pressure on life on Earth, then no more advanced than prokaryotic bacteria. Oxygen, indispensable to the survival of animals, was at that time a poison to perhaps the majority of life forms. Having evolved in environments with little or no oxygen, most life experienced the sudden appearance of the chemically reactive element as a global disaster—but for the rest it was a powerful evolutionary spur. There were but two choices facing life on Earth in that long-ago time: Adapt through evolution, or die.

All organisms in the sea had to adapt in two major ways. First, they had to evolve enzymes capable of mitigating the ravages of dissolved molecular oxygen and chemicals called hydroxyl radicals. (We humans are still trying to do this. Our ingestion of antioxidants such as vitamin E and vitamin C is an attempt to reduce the ravaging effects that dissolved oxygen and "free radicals" have on living cells.) Second, with the banded-iron formations' precipitation out from seawater, living cells no longer inhabited a solution rich in iron. After having been surrounded by high-iron solution since the first formation of life, proteins within cells had to be reengineered for life in an environment low in iron.

Recent DNA sequencing has shown that several enzymes found in archaeans and eukaryotes are left over from this event of 2.5 billion years ago. No such enzymes occurred in the older bacteria. The implications of this are profound: Kirschvink and his colleagues are proposing no less than complete rejection of the Tree of Life models we examined at the end of Chapter 3, which suggest that the three great domains (Archaea, Bacteria, and Eucarya) all arose soon after life's first evolution at least 3.8 billion years ago. The new study has not only uprooted this tree; it has burned it. If the Kirschvink group is correct, two of the three domains—Archaea and Eucarya—arose only *after* the 2.5-billion-year-old Snowball Earth and are thus much younger than the bacteria. Soon after this, in rocks about 2.1 billion years of age, we find a record of the oldest organelle-bearing eucaryan—the creatures known as *Grypania*, which we mentioned in Chapter 3.

This new version of the Tree of Life is a revolutionary scientific discovery, and if true, it will utterly reshape our understanding of life's evolutionary path. The Snowball Earth events can be seen as biologically important in two ways. First, the inception of the Snowball produced what may have been the largest "mass extinction" (the subject of Chapter 8) in our planet's history. The persistence of globally freezing temperatures, the isolation of the ocean from sunlight, the change in the precipitation patterns on Earth, and the removal of all water from the surfaces of continents would have removed the majority of surface habitats then available for microorganisms. In only a few places could microorganisms have survived: in the deep earth, around hot springs, and in hydrothermal deposits. Second, Earth's release from this icy prison after 30 million years brought about a new catastrophe: from cold to hot, from oxygen-free to oxygen-rich. Again, organisms had to adapt rapidly. It is this legacy that we may be seeing in the DNA of all living organisms; those that survived all bear witness in their DNA to this dual catastrophe— first cold, then warmth and oxygen. Life on the early Earth went through an icy bottleneck, and it came out the other side radically changed.

The Snowball Earth of 2.5 million years ago may have given our planet eucaryans and the eukaryotic cell necessary for animal life. The second series of Snowballs (there were several in rapid succession) may have bequeathed our planet an even more interesting biological legacy—animal life as we know it.

THE SECOND GLOBAL GLACIATION

As we saw in Chapter 5, by the next round of Snowball Earth events, those spanning the time interval from 800 to 600 million years ago, animal life was present on Earth, but it was newly formed. Either simultaneously with or soon after the appearance of the new animal phyla, Earth was once more locked into a global icehouse. Once again, there must have been a period of mass extinction, as the warm planet froze and the heat-loving organisms of

Earth had to retreat to oases of heat, such as around volcanoes and hydrothermal vents, or die. Yet the very severity of these events may have benefited the newly arisen animals. The great stress inflicted by environmental conditions imposed by the Snowball events would have stimulated inordinately rapid evolution among the newly evolved animals. It would also have caused the isolation of various populations, because the small populations of life huddled around the undersea volcanoes would have been cut off from any exchange of genes with other animal groups. This very isolation may have been largely responsible for the diversity of phyla that emerged at the other end of these crises, for when the final Snowball Earth event ended, about 600 million years (or less) ago, an entirely new group of creatures was ready to take over the planet. This is the interval when animal life began to diversify dramatically, in an event known as the Cambrian Explosion, the subject of the next chapter.

Would this have happened if the glaciations had not occurred? Kirschvink and Hoffman suggest that there is a causal link between the cessation of these major glaciations and the emergence of animals. Hoffman has noted, "Without these ice events, it is possible there wouldn't be any animals or higher plants." He believes that the melting of the ice at the end of these ice ages boosted biological productivity—and in the process stimulated evolutionary activity. This idea has yet to be confirmed, but it remains a tantalizing possibility.

Both of the two great episodes of Snowball Earth nearly ended life on Earth, as we know it. But each, ultimately, may have been crucial in stimulating the great biological breakthroughs necessary for animal life: the evolution of the eukaryotic cell and then the diversification of animal phyla. This leads us to ask whether Snowball Earth events are *necessary* to produce animal life as diverse as that seen on Earth today.

The end of the last Snowball Earth event brought the time interval known as the Precambrian to a close. Soon thereafter, abundant skeletons of larger animals began to fill the sea, in the Cambrian Explosion. If the two groups of scientists led by Joseph Kirschvink and Paul Hoffman are correct about Snowball Earth, a good case can be made that life on Earth is to some extent due to these events.

PLANETARY SURFACE TEMPERATURES AND THE EMERGENCE OF LIFE

The discovery of the Snowball Earth episodes suggests that temperature-induced events in planetary history may profoundly affect the course of biotic evolution. This argument can perhaps be extended not only to specific episodes of planetary temperature change but also to actual temperature values over time. Could cooling planetary surface temperature reaching some critical value have been the stimulus for other major breakthroughs in biological evolution?

As we saw in Chapter 2, the habitable zone is most commonly defined in terms of the presence of liquid water; this definition thus includes everything from life forms capable of living in boiling water to those capable of life in ice or snow. It may be that over much of its history, Earth was either too hot or too cold to allow the emergence of animals. Environments with temperatures near the freezing point or the boiling point of water are occupied largely by microbes; animals tolerate a much narrower temperature range. David Schwartzman and Steven Shore have pointed out that eukaryotic organisms with mitochondria (the organelles that convert fuel into energy) have an upper temperature limit for viable growth of 60°C. This limit is apparently determined by the chemical structure of the mitochondrial wall. Because eukaryotes evolved from prokaryotes, the habitable zone of a planet is narrowed from the region that allows the presence of water (0–100°C) to the narrower range of 0–60°C. Schwartzman and Shore note, "We assume that the emergence of relatively simple life forms is almost inevitable on Earth-like planets. Such organisms are remarkably robust. *Complex* life, however, requires a more restrictive set of physical conditions—in particular, lower temperatures."

Schwartzman and Shore provided the following list of the critical upper temperatures for various organisms on Earth.

Group	Approximate Upper Temperature Limit (°C)	Time of First Appearance on Earth (billions of years ago)
Multicellular plants	45–50	0.5
Animals	50	1–1.5
Eukaryotic microbes	60	2.1–2.8
Prokaryotic microbes		
Cyanobacteria	70–73	3.5
Methanogens	>100	3.8
Extreme thermophiles	>100	3.8

Schwartzman and others have proposed that Earth's surface temperatures have been the critical constraint on microbial evolution, determining the timing of major innovations. They believe that when Earth's surface cooled below 70°C more than 3.5 billion years ago, cyanobacteria were able to evolve. These microbes colonized the surface of the land and, in so doing, increased weathering rates and soil formation. The new soil in turn acted as a sink for removing carbon dioxide from the atmosphere, thereby causing further cooling of the planet. Each innovation among microbes resulted in biotic enhancement of weathering. This process has been dubbed "biotically mediated surface cooling." With the evolution of higher plants with their often elaborate root systems, this process increased greatly in efficiency. It is at the heart of the "Gaia Hypothesis," wherein Earth is conceived of as a self-regulating "superorganism," a perspective shared by many scientists, including Lynn Margulis, Tyler Volk, and the originator of the term, James Lovelock. We remain agnostic on this particular interpretation but see much merit in the view that the emergence of animals may have been strongly influenced by surface temperatures—and that life itself on this planet has had an enormous impact on planetary temperatures.

Could there be any way in which (or any planet whereon) animals could evolve faster than they did on Earth? The physical events affecting Earth immediately before the emergence of large, skeletonized animals were among the most complicated in all of Earth history. Was this just coincidence, or did it make the acceleration of animal evolution possible? These questions, and the curious and dramatic fashion in which the major body plans of animals suddenly began to commonly appear in the fossil record on our planet about 540 million years ago, are the topics of the next chapter.

The Enigma
of the
Cambrian
Explosion

Evolution on a large scale unfolds, like much of human
history, as a succession of dynasties.

—E.O. Wilson, *The Diversity of Life*

Our planet was without animal life for the first 3.5 billion years of its existence and was without animals large enough to leave a visible fossil record for nearly 4 billion years. But when, 550 million years ago, sizable and diverse animal life finally burst into the oceans, it did so with a figurative bang—in a relatively sudden event known as the Cambrian Explosion. Over a relatively short interval of time, all of the animal phyla (the categories of animal life characterized by unique body plans, such as arthropods, mollusks, and chordates) either evolved or first appear in the fossil record. Undoubted fossils of metazoan animals have never been found in 600-million-year-old sedimentary strata, no matter where on Earth we go.

Yet the fossils of such animals are both diverse and abundant in 500-million-year-old rocks, and they include representatives of most of the animal phyla still found on Earth. It appears that in a time interval lasting at most 100 million years (and in fact, as we will see, in an interval considerably shorter than that), our planet went from a place without animals that could be seen with the unaided eye to a planet teeming with invertebrate marine life rivaling in size almost any invertebrate species on Earth today. This follow-up to the initial animal diversification of more than 700 million years ago (described in the last chapter) is the Cambrian Explosion.

The rate of evolutionary innovation and new species formation during the Cambrian Explosion has never been equaled. The prior animal diversification must have involved very few species, each growing to a very small size; the Cambrian Explosion, on the other hand, produced huge numbers of new species, many with completely novel body plans. The Cambrian Explosion presents a great challenge to astrobiology, as we will show in this chapter. Questions abound. For instance, can there be animal life on a habitable planet *without* this type of event? Is the Cambrian Explosion an effect or a cause? That is, could it be that the remarkable animal diversity on Earth today is a by-product of this sudden diversification and would not have come about if the Cambrian event had been a mild bang rather than an explosion? Was it inevitable once the late Precambrian first event had occurred, or was another set of stimuli required? What animals were involved? What were the event's biological origins? What caused it to occur? (Was there some sort of biological or environmental trigger?) And, most relevant to astrobiology, was the Cambrian Explosion inevitable once a certain level of biological organization had evolved? In other words, is there any way that the Cambrian Explosion might *not* have occurred?

WHEN DID THE CAMBRIAN EXPLOSION TAKE PLACE?

The Cambrian Explosion is marked by the sudden appearance of larger fossils, which can be easily seen at many places around the globe. There is nothing subtle about this evidence, and it was known to even the earliest geologists. In Washington State, for instance, telltale signs of the Cambrian event are readily visible near the small town of Addy, where a slow country road meandering through the Coast Range foothills cuts through low outcrops of quartzite, the lithified remains of what, more than 550 million years ago, was a white, sandy beach. If we could travel back to that time, the beach itself would probably not elicit our wonder, for it would appear completely unremarkable. Beaches are beaches regardless of time. But the appearance of the nearby shore and inland vistas *would* be remarkable; there would be no plants (or animals) to be seen. On Earth today there are a few areas where plant life is not immediately visible—the harshest deserts, the Arctic and Antarctic regions—but these are exceptions on a planet otherwise carpeted with life. Yet such was not the case 550 million years ago. And it was not only the land that was barren: If we were able to wade through the shallow, warm sea, we would not encounter any shimmering fish or scuttling crabs, no starfish and no sea urchins. There would be no clams burrowed in the sand and almost none of the other animals we see so commonly along the seashores of our world. There might be a few worms or jellyfish, but nothing with a readily visible skeleton. We would conclude that this world boasted little life, or at least little that we would recognize as animals or land plants.

The quartzite in this region is exposed as sedimentary layers stacked one on top of another, and if we were to count the individual layers (or beds), they would number in the thousands. The lowermost beds are devoid of fossils. Yet because these rocks are stratified, they are organized by time. If we wander a short distance farther along the roadside outcrops, moving upward through the succession of these stratified beds (and thus into younger intervals of time preserved in the rocks), we see a wondrous thing. Suddenly, as though by magic, an abundance of fossils appears. We find the remains of shelled creatures called brachiopods, which look like small clams, and a few other types of fossils, such as sponges and a tiny mollusk or two. But by far the most common fossils to be found are also the most spectacular, for the first fossil-bearing beds at Addy are packed with trilobites.

Along with ammonites and dinosaurs, trilobites are perhaps the most iconic of all fossils. At first glance they seem to resemble large bugs or crabs of some kind, but upon closer examination they look like nothing still alive; their closest living relatives are horseshoe crabs and pill bugs, but these are only distant cousins. Trilobite fossils range in size from the microscopic to nearly 3 feet in length. They have numerous spines, great helmet-like heads, and a variety of peculiar eyes, and their undersides housed an array of legs, gills, and other assorted arthropod tools. All in all, they are complicated fossils from complicated creatures—and, for that reason, are unlikely candidates for the honor of being the world's oldest animal fossils. If Darwin's theory of evolution is correct, the first fossils should be far simpler than a trilobite—as indeed they are. Yet at Addy, as at so many other localities around the world with sedimentary rocks of this age, the first obvious fossils are indeed trilobites perched atop thick sequences of strata apparently devoid of fossils. This observation suggests that animals of staggering complexity appeared on Earth without evolutionary precursors. It is as though an orchestra began playing without sounding a single tone to tune up.

This sudden appearance of larger animals in the fossil record is the most dramatic aspect of the Cambrian Explosion. It drove Charles Darwin to distraction and challenged the newly evolving field of geology, which had taken as its guiding principle the idea that important events in Earth history unfolded gradually, not abruptly. Yet even to the earliest geologists, the Cambrian Explosion seemed anything but gradual.

In the early nineteenth century, geology was a newly born scientific discipline established largely for economic motives, such as the search for fuel and metals. It was clear that the discovery of these valuable commodities depended on finding the relative age of rocks. By that time it was also recognized that fossils were the remains of ancient life and that they appeared in a relative, superpositional order and thus could provide a practical and reliable method for determining the relative ages of rock bodies. With the aid of fossils, geologists soon began to subdivide Earth's sedimentary strata into time units.

In 1823, English geologist Adam Sedgewick named one such unit the Cambrian. Sedgewick observed that a thick sequence of sedimentary rocks in

Wales contained a characteristic assemblage of fossils, including numerous trilobites. Overlying these strata were sedimentary rocks with a different suite of fossils that represented a time unit eventually named the Ordovician. Yet as he continued mapping and describing the mineral and fossil content in his field area, Sedgewick encountered something novel: strata *without* fossils. The Welsh sedimentary rocks studied by Sedgewick were composed of an enormous thickness of unfossiliferous strata, overlain by an equally thick pile of strata containing trilobites and brachiopods. Even more curious, the transition between the unfossiliferous and fossiliferous strata was abrupt, not gradual.

The strata *beneath* the fossiliferous, Cambrian rocks became known as the Precambrian. The Cambrian period was defined as the block of time during which the fossil-bearing strata recognized by Sedgewick in Wales were deposited. Thanks to modern dating techniques, we now know that this unit of time started about 540 million years ago and ended about 490 million years ago. Although Sedgewick's strata are found only in a part of Wales, we refer to all rocks on Earth that formed between 540 and 490 million years ago as belonging to the Cambrian system.

Sedgewick defined the base of the Cambrian as the stratal level where the first trilobite fossils could be found, and that view prevailed for over a century. Anywhere in the world where trilobite-bearing strata overlay unfossiliferous strata was considered to mark the base of the Cambrian. Recently, however, the way in which the base of the Cambrian is recognized has changed. It is now marked at a level that Sedgewick would have considered below the "base" of the Cambrian. Today geologists use the first occurrence of a particular *trace fossil* (the fossilized record of animal behavior, rather than the preserved hard parts of the animal itself) as the base of the Cambrian system.

Sedgewick's discovery of the seemingly instantaneous appearance of complex fossils convinced most scientists of his time that life was spontaneously created—put on Earth through the action of some deity; this observation is still cited by creationists as evidence against the theory of evolution. This observation was perhaps the most difficult for Charles Darwin to reconcile with his newly proposed theory of evolution, for the apparently sudden appearance of large, complex animals in the fossil record ran utterly contrary to his expectations. In *On the Origin of Species*, he speculated that the Precambrian interval must have been of long duration and "swarmed with living creatures." Yet where were the fossils of these swarms? Surely, if Darwin was correct, a long period of evolutionary change with simpler precursors would have been necessary to produce the complex creatures collected by Sedgewick and others in the lowest strata now known as the Cambrian. Darwin was never able to refute this stringent criticism of his theory. Instead, he railed against the "imperfections" of the fossil record, believing that there must be a missing interval of strata just beneath the first trilobite-bearing beds everywhere on Earth. He was convinced that there must be Precambrian-aged fossils. As it turns out, he was right, but he went to his grave unvindicated.

Paleontologists have since proved Darwin correct, for the supposedly "barren" strata beneath those that bear what were thought to be the first fossil skeletons do indeed contain the ancestors Darwin sought and theorized about. They were long overlooked or missed, however, because of their rarity or very small size. Most organisms from the youngest "Precambrian" time both were tiny and lacked skeletons, so they rarely left obvious traces in the fossil record. They are very hard to detect unless special processing techniques are used to extract them from their entombing matrix; Darwin and his contemporaries had not yet dreamed of such methods. The supposedly "sudden" appearance of skeletonized life, more than 540 million years ago, is simply the first appearance of creatures with large skeletons, which produce fossils that are easily noticed. Because of this, the base of the Cambrian has now been "lowered" into the supposedly barren strata beneath the first trilobite-bearing beds. Just as Darwin supposed, trilobites do appear only after a longer period of evolution of simpler forms that rarely fossilize.

The twentieth century has witnessed a revolution in the science of geology. No longer are fossils the sole means of dating rock. Sophisticated laboratory analyses of volcanic and some sedimentary rock give accurate ages in years, and the entire rock record (including the Cambrian) has been far more accurately dated. In the 1960s the base of the Cambrian was determined to be 570 million years old, and this date appears on age compilations even into the late 1980s. Recently, however, there have been significant improvements in radiometric dating techniques. The Precambrian/Cambrian boundary is now dated at 543 million years old. The "Middle" Cambrian is dated at about 510 million years ago, whereas the oldest trilobites are no more than 522 million years old, which suggests that the bulk of Cambrian time was "pre-trilobite." Interestingly enough, although the "base" of the Cambrian has gotten younger; its "top" has not changed in age. The Cambrian Explosion remains a relatively sudden and signal outburst of animals— an unleashing of abundant and voracious creatures upon the earlier bacterial world, which continues, unabated, more than half a billion years later. With the exception of life's first formation, it remains the most profound biological event to have occurred on this planet. And we propose that the Cambrian Explosion has an even greater significance than Charles Darwin (or modern scholars of the fossil and evolution record) realized: We believe that it yields crucial evidence for estimating the frequency of animal life in the Universe.

WHAT ANIMALS WERE INVOLVED IN THE CAMBRIAN EXPLOSION?

No one disputes that a huge diversity of large animals emerged with alacrity between 600 and 500 million years ago. The event itself took place in the sea, for the land areas of the time were largely barren except for lichens and perhaps a few low plants; there were no trees, no shrubs, no stemmed plants at all. Because of the lack of rooted vegetation, little soil would cling to the land surfaces.

In the shallow seas and waterways, however, life was plentiful (though clearly different from that in the seas of today, as we noted above) and was rapidly changing in composition. Stromatolites, the layered bacterial forms that had been the dominant type of life on Earth for most of the 4-billion-year Precambrian era, were by 500 million years ago nearly absent from the planet. There were literally being eaten out of existence, for a great biological

131

revolution was creating entire suites of organisms adapted for utilizing plants as food. These newly evolved grazing animals (many looked like small worms) used the stromatolites as food. After 700 million years ago, a steep decline in stromatolite diversity took place, and newly evolved herbivores were surely its cause, although these grazing creatures left no fossil record. (They were too small and had no mineralized skeletons that could fossilize.) In most instances we simply infer their existence.

Thus the stage was set for the great evolutionary drama we call the Cambrian Explosion. It was grand theater, composed of four acts, each with its own set of characters, although some of them hung around for successive acts before exiting—by going extinct!

Act 1: The Ediacarans

The first act introduced a truly odd assemblage of creatures that looked like bizarre jellyfish, mutated worms, and quilted air mattresses somehow brought to life. This opening cast of characters is collectively known as the Ediacaran fauna.

We now know that the Ediacarans opened Act 1 about 580 million years ago and were largely gone by 550 million years ago (although a few appear in much younger rocks). Most of the Ediacaran fauna somewhat resemble members of the phyla Cnidaria and Ctenophorata—the jellyfish, sea anemones, and soft corals of our world. Two of the most common types of Ediacaran fossils resemble jellyfish and stalked, colonial sea anemone-like animals known as sea pens (still quite common in our world), and they were first interpreted as early versions of these modern forms. Other members of the fauna were more worm-like in appearance, but these were minor players.

In some cases these are large organisms—some have left fossils nearly 3 feet long, making them veritable behemoths for their time. Yet they seemed to have little organization of the sort we are so familiar with. For example, they had neither an observable mouth nor an anus. Their organization suggests a series of tube-like structures quilted together. In a 1988 essay, Stephen Jay Gould proposes that these odd animals are indeed the flowering of the "diploblastic," or two-cell-layer body plan, a type of body plan found today only in the corals and jellyfish.

The Ediacarans were not discovered until the 1940s, when an Australian geologist named R.C. Sprigg noticed some odd-looking fossil remains on scattered slabs of sandstone mines in the Ediacaran Hills of southern Australia, a desolate and isolated locality in very arid country. The fossils were simply impressions in the sandstone, rather than preserved skeletons of any sort. Some were worm-like; others looked like giant leaves; a third group were circular in shape. Sprigg collected a few of these, noting that many of the circular impressions made in the sandstone looked like modern-day jellyfish, also known as Cnidarians. But such soft creatures as jellyfish are preserved in rock only under the most extraordinary circumstances, and because of this, many wondered whether these were fossils at all. Nevertheless, Sprigg briefly announced his find in a scientific journal, described them as "among the oldest direct records of animals in the world," and noted that "they all appear to lack hard parts and to represent animals of very varied affinities." The fossils ranged in length from less than an inch to more than 40 inches. Other fossils from this region began to turn up (see Figure 7.1), and they eventually became the passion of Australian paleontologist Martin Glaessner. He created the first biological reconstructions of the odd Ediacaran fossils and made astute observations about the nature of the environment in which these organisms lived. Intensive study of the taxonomic affinities of this varied fauna soon followed.

Glaessner ultimately placed the entire Ediacaran fauna, as he called them, into known phyla, such as the Cnidaria, a phylum thought to be among the most primitive of all animals. To him, the Ediacarans thus represented the first flowering of the animals and belonged to taxonomic groups still present on Earth today. Using the tree analogy, Glaessner viewed his Ediacarans as "missing links" between the small, presumably simple ancestors of all animals, and the jellyfish and anemones still alive today. The Ediacaran world seemed to resemble a Cnidarian world, and this agreed nicely with most biologists' view of how the metazoan radiation may have unfolded, beginning with the most "primitive" of phyla, the sponges and Cnidarians, followed only later by more complex fauna such as arthropods (and the trilobites, which are members

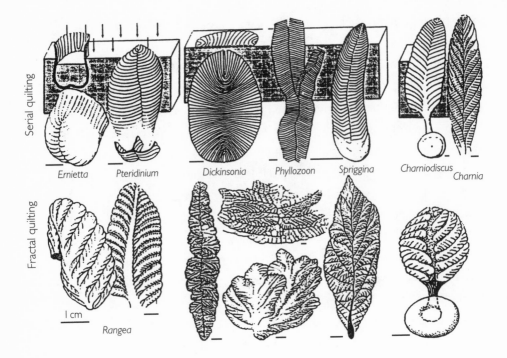

Serial quilting

Fractal quilting

Ernietta   *Pteridinium*   *Dickinsonia*   *Phyllozoon*   *Spriggina*   *Charniodiscus*   *Charnia*

1 cm

Rangea

Figure 7.1   *Lifestyles of bilateral Vendobionta. Ediacaran fauna. Drawings by A. Seilacher.*

of the phylum Arthropoda). According to this view, our modern-day animals are descendants of the Ediacarans. This opinion is still held today by many specialists, including Simon Conway Morris of Cambridge University.

In the nearly four decades since Glaessner's interpretations were first published, the Ediacaran fauna has achieved new importance. First, the strange fossils making up this assemblage have been found beyond Australia. In the White Sea region of Russia, Arctic Siberia, Newfoundland in Canada, and Namibia in southern Africa, other fossilized examples of these strange creatures are preserved, showing that the Ediacaran fauna was essentially worldwide in distribution at the end of the Precambrian. (Many of these localities have now been dated via radiometric geochronometry; some of the oldest known on Earth are those at Mistaken Point, in Newfoundland, dated as 565 million years old.) Second, the Ediacarans seem to have a longer stratigraphic range than previously supposed, and in a few places on Earth, they may actually have coexisted briefly with the undoubted animal faunas. Finally, some workers believe that the Ediacaran fauna do not represent animals at all but are large plants, fungi, or even lichens. Others class them as animals, but animals belonging to taxonomic groups now extinct. The Ediacarans have thus gone from logical precursors of the Cambrian Explosion to much more controversial players in the evolutionary drama.

This latter view, that the Ediacarans are not a main branch of the tree leading to animals, but rather represent a side branch now extinct (and thus have nothing to do with the ancestry of all current animal life), has been most effectively championed by paleontologist Dolf Seilacher of Yale and Tubingen University. He suggests that the resemblance between the Ediacarans and living creatures such as jellyfish and sea pens is coincidental. In his view, the Ediacarans represent an extinct assemblage of organisms—a separate biological "experiment" involving creatures with tough outer walls and fluid-filled interiors. Seilacher has speculated that during the time of the Ediacarans, a thick mat of bacteria covered the ocean bottoms, and this could answer the very perplexing question of how organisms without hard parts became fossilized.

This bacterial mat may explain why the soft-bodied Ediacaran fossils were so commonly preserved as fossils. As sand settled over the Ediacarans, they were pushed downward into the bacterial mat. Their tough outer walls could not be crushed readily, and the impressions they made in the mats were preserved in three dimensions by the overlying sand. Seilacher suggests that the evolution of new, efficient grazing animals such as mollusks at the start of the Cambrian rapidly brought an end to such mats and changed the way sediment accumulated in the earliest Cambrian.

Perhaps the most intriguing aspect of the Ediacarans is that there is no evidence of predation on them; we have no records of Ediacarans preserved with bite marks or missing pieces. Did these creatures live in a time without predators—in a "Garden of Ediacara," as paleontologist Mark McMenamin has dubbed the time?

What actually happened to the Ediacarans? Conway Morris asks whether they were diluted out of existence. That is, did the great radiation of

135

animals at the base of the Cambrian simply overwhelm them in the fossil record? (In other words, they existed but were too few to fossilize.) Was the disappearance of the Ediacaran body form a result of the first mass extinction on planet Earth? Were they ecologically replaced or even preyed on by the new animals—driven to extinction by efficient new predators against which they had little defense? The fossil record is still enigmatic. In some places on Earth, the Ediacarans are gone before the first "Cambrian" animals appear, which suggests that the new animals were simply filling the niches of the extinct Ediacarans. But as we have said, in other places there is a clear overlap between the two, suggesting that a competitive interaction ensued.

The Ediacarans do make good theater—enigmatic, mysterious, and the first on the stage. They were a hard act to follow, but a great wave of diversification was occurring at the end of their time in the limelight, a wave that continues on planet Earth still. With the second act of the Cambrian Explosion, undoubted animals appear on the scene.

Acts 2 and 3: Trace Fossils and Small Shellys

We can combine the next two acts, because the cast of characters is both incomplete and poorly characterized. In Act 2, a new group of players, seemingly wearing masks to disguise their true identity, replace most of our opening troupe. We detect them only by the footprints they left on the stage itself, for we have no true "body" fossils (usually the remains of skeletal hard parts). The second assemblage of life making up the Cambrian Explosion has left only squiggles and tracks in the ancient sediment. Such fossilized remains are known as trace fossils; they are not the *remains* of animals, but evidence of their behavior, and thus record the trackways or feeding patterns of ancient organisms. Yet they are of enormous significance. Whereas the Ediacarans simply sat in place their entire lives, these first trace fossils tell us that large animals capable of locomotion had appeared on Earth. Perhaps they were large worms or flatworms, or perhaps they belonged to phyla now extinct. The first and most primitive trace fossils appear in rocks as old as the Ediacarans, but they diversify and take center stage in younger rocks. Trace fossils are still being formed today and have been common in the rock record since the Cambrian. But they are clearly formed by many different organisms, and it is doubtful that the organisms that formed the first trace fossils survived into much more recent times than the Cambrian period itself.

Our Act 3 introduces an assortment of tiny calcareous tubes, knobs, and twisted spines, none larger than about ½ inch, all coming from animals that it is still impossible to reconstruct completely. Some are the remains of larger skeletons that have been fragmented into pieces, but most are single elements of some sort of a multielement skeleton, like individual spines coming from a porcupine. Collectively, they are known as small shelly fossils, or SSFs. The small shellys are first found in rocks dated to around 545 million years ago. These extremely significant fossils tell us that another great biological breakthrough had been achieved: The SSFs are the first large animals with mineralized skeletons.

Act 4: The Trilobite Faunas

Act 4 of our play is a grand finale featuring fossil icons much more familiar to us than the previous actors. They include the first trilobites, brachiopods, and a host of newly evolved mollusks and echinoderms. The characters are now far larger—and greater in number—than in any of the three previous acts, and ironically, these actors were long thought to mark the start of the Cambrian Explosion, rather than its end. This last group didn't appear until about 530 million years ago. Its diversification proceeded for another 30 million years. By about 500 million years ago, the Cambrian Explosion was finished.

The trilobites are by far the most diverse and obvious part of this assemblage. The oldest trilobites, of which the genus *Olenellus* is diagnostic, were spiny, somewhat resembled annelid worms, and had large crescent-shaped eyes. They all had walking legs and gills, and all appear to have fed by ingesting sediment or particulate material on the sea floor. They showed little adaptation for defense against predation.

Another curious group to appear contemporaneously with the trilobites were immobile, coral-like animals called archeocyathids. This group had conical skeletons made of lime and lived gregariously. They appear to have been the world's first reef-forming organisms and seem to have lived in the same environments favored by corals today. In addition to being the first of a long line of reef formers, the archeocyathids have another somewhat dubious claim to fame: They were perhaps the first animal phylum to go extinct. The basic body plan of the archeocyathid skeleton is unlike anything alive today. Taxonomists place these creatures in the same phylum as the living sponges, but this is as much for convenience as anything else. They appear to have constituted a separate phylum—and one of the few phyla we know of to have suffered utter extinction.

The remarkable Burgess Shale fauna of British Columbia has yielded extraordinary insights into animals living among the trilobites. Because of the lack of oxygen in this ancient environment, even soft parts were preserved, and these remains offer us an unparalleled window into the past. The Burgess Shale reveals how diverse the marine ecosystems were by the time trilobites evolved. Yet by the time of the Burgess fauna, some 505 million years ago, the majority of animal phyla appear to have been present.

WAS THE CAMBRIAN EXPLOSION INEVITABLE?

Darwin's theory of evolution describes two of the most important scientific discoveries ever made: (1) that all life has descended from a single common ancestor, and (2) that the various species descending from this ancestral creature have descended with modification. The great advances of physics and chemistry are milestones in human understanding, but they do not themselves describe life. We are life, and we have appeared on this planet through the processes of evolution; it is a central law affecting us. Yet for all its importance, the theory of evolution remains one of the most misunderstood of scientific views. One popular misconception equates evolution with increasing complexity and assumes that evolutionary change (Darwin's "descent with modification") always means an unbroken series of ever more complex organisms or structures within organisms. Although greater complexity often does evolve, it is not an end result of the evolutionary process; modification can occur without increases (or decreases) in complexity. We have only to look at the domains Archaea and Bacteria to see that this is true. From what we can tell from their fossil record, the archaeans and bacteria are no more morphologically complex now than they were 3.5 billion years ago (although, as we have noted, their biochemistry has diversified almost endlessly). They have evolved, to be sure, but that evolution has not involved dramatic increases in morphological complexity.

Of the three domains of life, only one, Eucarya, has undertaken wholesale experimentation in new morphology and body plans. If the process of life's creation were to be repeated innumerable times, it is not at all certain that eucaryan equivalents (lineages exploiting the morphological route of adaptation, rather than the chemical route utilized by the archaeans and bacteria) would appear each time—or even ever again. But on *this* planet, at least, the eucaryans *did* arise, and it was from this group that the multicellular animals now dominating planet Earth arose. The pattern and timing of their evolution on Earth may provide major clues to understanding whether, and how often, equivalents of our planet's complex animals could have arisen on other planets.

There are important astrobiological implications in this: Will animal life (or some other type of complex life) inevitably develop on all worlds in a planetary habitable zone? In our estimation, it has always been assumed that forming the first life was the hardest aspect, but that once life originated, it inevitably proceeded "up" gradients of complexity, culminating in very complex animals. Yet the actual history of life on this planet tells a different story. The first life appeared about 4 billion years ago. Eukarytotic organisms did not appear for another 1.5 billion years, and multicellular animals did not appear until more than 3 billion years after the first life. On the basis of this information alone, we would have to conclude that forming *animal* life is a much more difficult—or at least a more time-consuming—project than the initial formation of nonanimal life. Perhaps the timing observed on Earth was just chance; perhaps on any number of other Earth-like planets with newly evolved prokaryotic equivalents, animals would appear not billions, but millions, of years after life originated. Abundant evidence from our planet's history casts doubt on this possibility, however.

On Earth it is clear that the evolution of animals occurred not as a gradual process but as a series of long periods of little change, punctuated by great advances. This pattern of evolutionary "thresholds" was succinctly described by paleontologists Douglas Erwin, James Valentine, and David Jablonski in a 1997 article in *American Scientist:* "The fossil record of the last 3.5 billion years shows not a gradual accumulation of biological forms, but a relatively abrupt transition from body plans of single cells to those of a rich diversity of animal phyla." Evolution thus did not gradually create complex metazoans. They evolved quickly, probably in response to a set of environmental conditions quite different from those that allowed the evolution of life in the first place.

There were several of these "great leaps forward." One was the evolution of the eukaryotic cell type with its enclosed nucleus; another was the initial radiation of the animal phyla, described in the last chapter. The most profound, however, was the Cambrian Explosion, that short burst of evolutionary innovation that resulted in the appearance of the larger, complex animals we believe to be so rare in the Universe. In this single, approximately 40-million-year interval, all major animal phyla (all of the basic body plans found on our planet) appeared, each represented by some number of species.

This event has profound implications for the possibility of life on other planets. Is the pattern on Earth—a single, short-lived diversification of larger animals—unique or the standard for all planets? And why did the Cambrian Explosion not take place until 3 billion years after life's first appearance on our planet? Does evolution always require 3 billion years to transform a bacterium into a multicellular animal, or was evolution simply waiting for the environment to become conducive to the proliferation of animal life? This may be one of the most critical questions facing the emerging field of astrobiology.

The Cambrian Explosion signaled a major change in the *tempo* of evolution then prevailing on Earth. Prior to this, our planet's most complex life

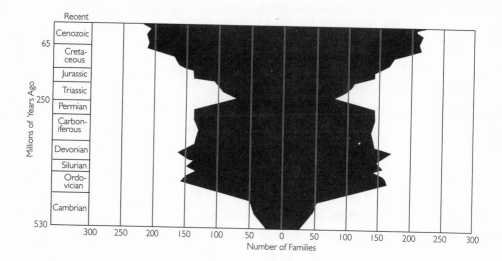

Figure 7.2   *Fluctuations in the number of families, and hence in the level of diversity, of well-skeletonized invertebrates living on the world's continental shelves during the past 530 million years are plotted by geological epoch in this graph. Time proceeds upward.*

consisted of algae, slime molds, and single-celled animals characterized by low rates of evolutionary change. There was little morphological change, and few new species arose over vast stretches of time (see Figure 7.2). The first evolution of metazoans changed all this. The staid tempo of evolutionary change that had characterized the first 3.5 billion years of life's history shifted into a higher gear. New species appeared at a far more rapid rate. They—*we*—have been diversifying at breakneck speed ever since.

A study of the various animal phyla is thus the study of a few dozen stable and long-lived body plans. This study has resulted in three great surprises. The first was the recognition that evolution has produced only a relatively few body plans. The discovery that the perhaps tens of millions of animal species on Earth today belong to between 28 and 35 phyla was a major surprise to nineteenth- and twentieth-century paleontologists and zoologists. Why this number and not a hundred? Or a thousand? Or five, for that matter? Diversification of the huge number of different species on Earth today (it is estimated to be between 6 and 30 million) has been through elaboration

141

or convolution of simple and conservative structural designs. Astrobiologists seek to discover whether this is how all animals (or their equivalents) evolve. Or is this simply Earth's way, and might there be worlds in space where there are nearly as many body plans as there are species?

A second surprise, and perhaps the most astounding, was that virtually all of the phyla appear to have originated no later than the end of the Cambrian and that *none* have appeared since. This cannot be proved, for there are some minor phyla (such as the rotifers) that have left no fossil record, and perhaps some phyla originated after the Cambrian. But there are none that we know of. For all the great changes that have occurred in the last 500 million years, with all the evolutionary events and mass extinctions of that long history, it would seem that at least a few new body plans would have appeared. Yet the fact that every phylum with a fossil record is represented in Cambrian strata makes such a supposition problematic.

The third surprise was that there may have been far more phyla on Earth in the Cambrian than there are today. Fewer than 40 extant animal phyla are recognized today. Yet according to some paleontologists, in the Cambrian that number may have been as high as 100! Although the number of species on the Tree of Life has been increasing through time, the number of higher taxa, such as phyla, has been *decreasing*. Thus the tree keeps adding ever more twigs and leaves on a dwindling number of major branches. Perhaps the Tree of Life on some other planet is quite different, with great new branches appearing continually through time.

WHAT—IF ANYTHING—TRIGGERED THE CAMBRIAN EXPLOSION?

Near the end of the last chapter, we pondered whether the initial diversification of the animal phyla was stimulated by evolutionary or environmental causes—specifically, the Snowball Earth events of between 800 and 600 million years ago. This same question can be posed about the subsequent Cambrian Explosion: Did it occur as late as it did in Earth history because it took that long for the *establishment* of an environment conducive to animals of large size, most with skeletons, or because it took that long for the necessary *genes* to evolve—genes that allowed the diversification of these metazoans? Genes or environment? Much new research into environmental conditions before and during the Cambrian Explosion, and into the physiological, anatomical, and genetic innovations leading to larger multicellular animals, has given us new views of this crucial moment in Earth history.

Many hypotheses have been proposed to account for the Cambrian Explosion. These can be categorized as attributing it to environmental causes or to biological causes.

Environmental Causes

• *Oxygen reached some critical threshold value.*

This is perhaps the most often discussed and widely favored of all environmental hypotheses. According to this hypothesis, the amount of available oxygen reached some critical level, or threshold, that made possible the great diversification of new organisms. Presumably, this level was higher than that during the first animal diversification event of 700 million years ago. Many scientists suggest that the biological response to a new, higher oxygen level was a biochemical breakthrough allowing animal life to construct hard skeletons for the first time. In the absence of abundant oxygen, organisms have a great deal of difficulty precipitating minerals as skeletal structures. As early as 1981, Heinz Lowenstam and Lynn Margulis postulated that skeletons in the form of collagen—an elastic, proteinaceous material similar to human fingernails—could have appeared as early as 2 billion years ago, for collagen formation does not require as much oxygen. Calcareous and siliceous skeletons and shells, however, were not possible at that early date.

• *Nutrients became available in large amounts.*

Just as a lawn needs fertilizer, ecosystems—and especially marine ecosystems—need a supply of organic and inorganic nutrients to remain at high levels of productivity and diversity. Abundant evidence suggests that the late Precambrian interval witnessed a relatively sudden and dramatic increase in nutrients, which may have had a significant effect on the evolution of organisms.

One of the mineral types most commonly found in rocks of this age is phosphorite, a mineral rich in phosphorus, one of the most important of the inorganic nutrients necessary for life. (The others are nitrate and iron.) There appears to have been a long interval during the late Precambrian when phosphates and nitrates were unavailable to organisms of the time, because they were buried in deep-water bottom sediments. However, the latest Precambrian was a time of changing oceanographic conditions; episodes of upwelling became common, in which deep waters were brought up to the sea surface, in the process liberating nutrients formerly locked in bottom sediment. This upwelling appears to be related to changing continental configurations.

The latest Precambrian was a time of intense plate tectonic activity. In particular, a giant "supercontinent" named Rodinia began to tear apart, and as this occurred, it changed the global patterns of ocean current circulation, thereby triggering the upwelling. According to this hypothesis, the release of the phosphate nutrients accompanying that new tectonic activity sparked the Cambrian Explosion.

• *Temperatures moderated following the late Precambrian "Snowball Earth" events.*

Just as the evolution of humanity is set against the backdrop of a global ice age, so too is the Cambrian Explosion associated with glaciation, for it occurred soon after the cessation of the "Snowball Earth" events profiled in the last chapter. As the last of these glaciation events finally ended, it signaled a protracted warming of the planet that followed 200 million years of glacial advances and retreats. Was this the trigger that unleashed the Cambrian Explosion, as suggested in the last chapter?

• *The Inertial Interchange Event*

There is a final environmental possibility that borders on the fantastic—as well as the believable. For many years, paleomagnetists have known that most, if not all, of the continents underwent large amounts of continental drift during Cambrian time. As we will see in more detail in Chapter 9, continental positions have an extraordinary effect on global climate, often controlling where warm and cold currents flow, the formation of ice caps, and even the abundance of greenhouse gases in the atmosphere. During the past few years, advances in the numerical calibration of the Cambrian time scale and improvements in the paleomagnetic database have revealed something astounding: Much of this continental drift happened during the Cambrian evolutionary explosion, and the entire episode lasted no more than 10 to 15 million years. The continental shifts were quite dramatic. North America moved from a position near the south pole to the equator, and at the same time, the entire supercontinent of Gondwanaland spun around a point in Antarctica, sending North Africa from the pole to the equator as well. It is as though the continents suddenly became ice skaters, gliding about Earth's surface with unprecedented ease for a short period of time, before turning to stone once more.

In 1997 the ubiquitous Joseph Kirschvink and two colleagues published, in the prestigious journal *Science*, a controversial interpretation of the cause of this tectonic movement—an explanation that is either the harbinger of a revolution in our understanding of planets and their histories, or unmitigated balderdash. As of this writing, the scientific community is about evenly divided and is awaiting further developments with keen anticipation. Kirschvink, David Evans, and Robert Ripperdan proposed that the Cambrian Explosion might have been triggered by another unique event in Earth history: a 90-degree change in the direction of Earth's spin axis relative to the continents. Regions that were previously at the north and south poles were relocated to the equator, and two formerly equatorial positions on opposite sides of the globe became the new north and south poles. This interesting hypothesis will be confirmed or discredited only through the acquisition of much new paleomagnetic data.

Kirschvink and his colleagues noted that all of the world's continents experienced a major increase in continental plate motions (the "drift" movement of continental drift) during the same short interval of time when the great evolutionary diversification took place—between 600 and 500 million years ago. This rapid movement of Earth's upper surface relative to its interior is thought to have been brought about by an imbalance in the mass distribution of the planet itself. During this redistribution, the theory goes, all the *solid* portions of Earth move together. But because Earth also has liquid portions (its inner core, for instance), the outer layer essentially flips over relative to the spin of the planet. This phenomenon would not be limited to Earth; it may also have occurred on Mars. Kirschvink and his colleagues point to the large volcano Tharsis, now located on the Martian equator. Tharsis sits atop the largest gravity anomaly (a center of high mass that creates forces of gravity greater than those in the surrounding rock) known for any planet in the solar system—a place of such high density that it creates a measurable perturbation in the planet's gravitational field. It is unlikely that Tharsis formed on the equator, according to Kirschvink and his colleagues. They believe the law of conservation of mass caused it to migrate later to its current equatorial position. Its movement would have been caused by an "inertial interchange event" similar to that posited for the Cambrian Earth. Once the volcano was at the equator, Mars would rotate so that its maximum moment of inertia was aligned with the spin axis.

The Earth's own inertial interchange event (IIE) would have taken only about 15 million years, and it is the very rapidity of this movement that causes Kirschvink and his colleagues to speculate that the IIE might have been associated with the Cambrian Explosion of life. During this period, existing life forms would have had to cope with rapidly changing climatic conditions, such as polar regions sliding to the hotter equatorial zones and warmer, low-latitude sites moving upward into the high-latitude, cold regions of the planet. These motions would have disrupted oceanic circulation patterns and would have perturbed most ecosystems on Earth. It might be only by chance that a unique tectonic event during Earth's 4.5-billion-year history coincided with a unique biological event. But how often does someone win million-dollar lotteries on two successive days?

The inertial interchange event can also explain one of the most curious aspects of Earth at this time. It is well known among geologists that the late

Precambrian and earliest Cambrian Earth underwent some sort of event that is preserved as large swings in carbon isotopes. (These are chemical signals found in the oceans in response to varying amounts of life on the planet; such signals were used to detect the first life on Earth in the Isua strata of Greenland, as described in Chapter 3.) About a dozen of these swings occurred near the end of the Precambrian time interval, and they have long puzzled geologists. The swings in these isotopes suggest that large amounts of organic carbon, long buried in ocean sediment, were suddenly exhumed and reintroduced into Earth's carbon budget. Repeated, major changes in the oceanic circulation patterns could produce these effects, yet such global changes would require massive tectonic changes in short periods of time. These changes would have fragmented ecosystems and could have prompted evolutionary diversification. The inertial interchange event would accomplish this.

If the Cambrian Explosion was necessary for animals to become so diverse on this planet, and *if* the inertial interchange event occurred as postulated, and *if* the Cambrian IIE event contributed to the Cambrian Explosion or even somehow was required for the Cambrian Explosion to take place, then Earth as a habitat for diverse animal life is rare indeed.

Biological Causes

In his pivotal book *Oases in Space,* paleontologist Preston Cloud suggested that there were four biological prerequisites for the Cambrian diversification event to take place: the prior presence of life itself, the attainment of oxidative metabolism (the ability to live and grow in the presence of oxygen), the evolution of sex in the domain Eucarya, and the presence of an appropriate protozoan ancestor to give rise to more complex animals. In Cloud's view, attaining all of these milestones took nearly 4 billion years—85% of Earth's history. He thus seems to believe that biological actors were more important in creating the Cambrian Explosion event than were the environmental aspects we considered in the previous section. But other biological factors must have played a pivotal role as well.

- *The advent of precipitated skeletons*

Skeletons are critical to large body size for many animals. Skeletons usually perform several functions, such as protection (from predation, desiccation, and ultraviolet rays), muscle attachment (thus allowing locomotion), and maintenance of body form. Yet building such structures required many evolutionary breakthroughs. Oxygen levels would have been critical for two reasons. First, large skeletons such as a shell covering (found in the earliest trilobites and mollusks) restrict the access of seawater to the soft body parts. In most early animals, respiration took place by direct adsorption of oxygen from seawater, across the body wall. Second, the presence of a shell means that a larger area of the body is no longer available for this type of respiration. In low-oxygen conditions, animals have a difficult time getting enough oxygen as it is, and adding a body cover only makes this problem worse. Thus, skeletons such as shells would not have evolved until relatively high oxygen levels were available in seawater.

Professor Dolf Seilacher (whom we met in our discussion of the Ediacaran fauna) is convinced that acquisition of skeletons played the dominant role in causing the sudden appearance of multicellular animal phyla. He notes that hard skeletons are not simply additions to preexisting body plans. Their very evolution *modifies* body plans. Seilacher argues that the Cambrian Explosion was triggered not by environmental conditions that allowed larger animals to develop but by those factors that allowed skeletons to appear. This is a subtle but important distinction. With the ability to produce hard parts, new animal groups could use these hard parts for jaws, legs, or body support, and this enabled them to exploit entirely new ways of life and new environments.

- *Attainment of evolutionary thresholds made large size possible.*

A second possibility is that evolutionary breakthroughs allowed, for the first time, the advent of large body size. We know that the majority of living organisms up until this time were less than a millimeter long; most were far smaller. Did genetic innovations allow larger body sizes and thus trigger the Cambrian event? Examples of such innovations include more efficient organ systems, such as improved circulatory, respiratory, and excretory systems. Each had to evolve before larger body size was attainable.

- *The predation hypothesis*

In 1972 paleontologist Steven Stanley (and later Mark McMenamin) proposed that the evolution of predators played a part in stimulating the Cambrian Explosion. Survival was enhanced in those animals that evolved the ability to defend themselves against predators by producing shells, burrowing deeply, or swimming or otherwise rapidly moving away from danger. And these creatures *incidentally* found themselves in a position to exploit food resources that had been underutilized or had not been utilized at all during the Precambrian. The evolution of shells made possible new forms of filter-feeding, and deep burrowing gave these animals access to new food resources. Cambrian predators thus forced animals to undertake new lifestyles, which turned out to be successful.

IS THE CAMBRIAN EXPLOSION SIMPLY AN ARTIFACT OF THE FOSSIL RECORD?

In the simplest sense, the Cambrian Explosion was a relatively sudden proliferation of animal types. The number of new species involved in this event is unknown, but it was several thousand at most and perhaps far less than this. The extraordinary aspect of this event was that the new species were spread among many new body plans. As we have said, each body plan defines a higher taxonomic category, such as a phylum or class. The Cambrian Explosion thus involved a large number of higher taxa, each of which was composed of just a few species. But are we simply seeing the advent of effective fossilization, rather than a real diversification "event"?

We recognize the Cambrian Explosion as such for a simple reason: We see a large number of fossils suddenly appear in the fossil record. But are we seeing a genuine flowering of new forms, or do the fossils merely mark the first appearance of *skeletons* in groups that had already been long established?

In other words, is the Cambrian Explosion merely an artifact of a very imperfect fossil record? Skeletons make fossilization possible; it may be that the actual diversification of body plans that appears to mark the Cambrian Explosion actually took place long before but is invisible to us because it took place among small animals without skeletons, which left no fossils.

This latter view can be considered the null hypothesis. What if there really wasn't a Cambrian Explosion at all? It may be that the various animal phyla accumulated in gradual fashion over the last billion years of the Precambrian, evolving one from another, but doing so without leaving any identifiable fossil record. It was thus only the evolution of large size, and of skeletons allowing the preservation of fossils, that accounts for the "Cambrian Explosion."

Whether the Cambrian event included the diversification of body plans or consisted simply of the first evolution, by these various body plans, of skeletons and large size is a moot point. *Something* stimulated the evolution of many large animals with skeletons in a brief period of geological time. Furthermore, M. McMenamin and R. McMenamin, in their 1990 book *The Emergence of Animals,* emphasize that mineralized skeletons—and especially shells—profoundly influenced the evolution of new body plans. A wide variety of forms use their shells not only for protection but also as an integral part of feeding. Brachiopods and bivalves (both invertebrates with two shells) use the shell as an integral part of a filter-feeding process. It is hard to see how the basic body plan of each of these groups could have formed before shells did.

CAMBRIAN EXPLOSION, CAMBRIAN CESSATION

For all of the animal phyla to appear in one single, short burst of diversification is not an obviously predictable outcome of evolution. Although there was certainly a long (200-million-year?) interval from the appearance of the first metazoans until the 20- to 30-million-year Cambrian Explosion, most of their morphological diversification during this time, including acquisition of the skeletal parts so characteristic and diagnostic of many invertebrate phyla, took place in a relatively short period.

Yet as wondrous and unexpected as that finding is, a second aspect of the evolution of the phyla is equally puzzling. The Cambrian Explosion marked not only the *start* of the majority of phyla as recognized in the fossil record but also the *end* of evolutionary innovation at the phyla level: Since the Cambrian, not a single new phylum has evolved. The extraordinary fact is that the diversification of new animal body plans started and ended during the Cambrian Period. Is this evolutionary pattern a characteristic of animal life on all (or any?) planets that succeed in producing animals, or is it unique to Earth?

The lack of new phyla and the paucity of new classes after the end of the Cambrian Explosion may again be an artifact of the fossil record; perhaps many new higher taxa did evolve and subsequently went extinct. This seems unlikely. It is far more likely that the great surge of innovation that marked the Cambrian came to an end as most ecological niches became occupied by the legions of newly evolved marine invertebrates.

Yet a puzzling mystery remains: Subsequent to the Cambrian explosion, Earth suffered several major mass extinction events—short periods when a majority of the species then living on Earth went extinct. These events, profiled in detail in the next chapter, drastically reduced diversity. The most catastrophic of these, the Permo-Triassic mass extinction of 250 million years ago, eliminated an estimated 90% of marine invertebrate species, and thus provides a natural experiment that we can examine to understand better those factors that caused the Cambrian Explosion. And what we observe is that even after this major reduction in diversity, no new phyla appeared. Although the number of species plummeted to levels similar to the very low species diversity found early in the Cambrian, the subsequent diversification in the lower Mesozoic involved the formation of many new species, but very few higher taxonomic categories. The evolutionary events during the Cambrian and the Early Triassic are dramatically different. Both produced myriad new species, but the Cambrian event resulted in the formation of many new body plans, whereas the Triassic event resulted only in the formation of new species exhibiting body plans already well established.

Two hypotheses have been proposed to explain this significant differ-ence. The first supposes that evolutionary novelty comes about when eco-logical opportunities are truly large. During the Cambrian, for instance, there were many habitats and resources that had not been occupied or exploited by marine invertebrate animals, and the great evolutionary burst of new body plans was a response to these opportunities. This situation was not duplicated after the Permo-Triassic mass extinction. Even though most species were ex-terminated in this catastrophic event, enough representatives of various body forms survived to inhabit most of the available ecological niches (even if at low diversity or abundance) and, in the process, to discourage evolutionary novelty.

The second possibility is that new phyla did not appear after the Permo-Triassic extinction because the genomes of the survivors had changed enough since the early Cambrian to inhibit wholesale innovation. In this sce-nario the evolutionary opportunities were available, but evolution was unable to create radically new designs from the available DNA. This is a sobering hypothesis and one not easily discredited, for we have nothing to which to compare the DNA we find in living animals. It could be that genomes gradu-ally become encumbered with ever more information—they gather more and more genes—and in the process become less susceptible to a critical mutation that could open up the way to innovation.

DIVERSITY AND DISPARITY

One of the central (and controversial) aspects of the Cambrian Explosion— especially with reference to the wondrous assemblage of fossils found in the Burgess Shale localities in Western Canada (where not only early animals with hard parts but also forms without skeletons are preserved as smears on the rocks)—concerns what are called diversity and disparity. *Diversity* (or in this case *biological diversity*) is a term familiar to most of us. Overtly, it is usually understood as a measure of the number of species present. Biologists use a more sophisticated meaning, that encompasses not only the number of species present in a locality, but also the relative abundance of those species. For instance, in this more technical sense, an assemblage of organisms comprising some given number of species, each with the same number of individuals in each species, is considered more diverse than a second assemblage with the same number of species but a highly unequal distribution of the numbers of individuals that compose each species group. *Disparity* is a measure of the number of body plans, types, or design forms, rather than the number of species. This distinction, first articulated by paleontologist Bruce Runnegar, seems at first glance rather subtle. Surely each different species has a somewhat different body plan from every other, and thus disparity and diversity should always be equal. But this is not the case. There are millions of species on Earth today. Yet the number of general body plans is far less than this.

Among animals, the major body plans are found within the major evolutionary lineages, the phyla. As we have seen, these animal groups all arose in the Cambrian Explosion. Yet the surprising finding of paleontology is that there were very few species in the Cambrian. In his 1989 book *Wonderful Life*, Stephen Jay Gould describes this finding as "a central paradox of early life: How could so much disparity in body plans evolve in the apparent absence of substantial diversity in number of species?"

The history of diversity and disparity during the Cambrian Explosion (or, more properly, *creating* the Cambrian Explosion) is another puzzling aspect of planet Earth's diversification of animals: Is this the only way to create animals, or just one way? Will every planet with animals create them as ours did, by evolving the entire spectrum of body plans in one great evolutionary rush among a low number of species? Or could this process be more gradual, with a slowly increasing number of species over long periods of time incrementally enlarging the number of body plans?

The Burgess Shale is clearly of major importance in understanding the initial diversification of animal life. It is largely responsible for showing us that most or all of the various animal phyla (or major body plans) originated relatively quickly during the Cambrian. But the Burgess Shale may also be telling us that not only were the body plans found on Earth today around in the Cambrian, but so too were other body plan types now extinct. One of the central messages of Gould's *Wonderful Life* is that the Cambrian was a time not only of great origination but also of great extinction, for Gould (and others as well) assert that far more phyla were present in the Cambrian than exist today. How many were there? Some paleontologists have speculated that there may have been as many as 100 different phyla in the Cambrian, compared to the 35 still living today. Gould clearly believes that there were more Cambrian than present-day phyla: "[W]e may acknowledge a central and surprising fact of life's history—marked decrease in disparity followed by an outstanding increase in diversity within the few surviving designs."

This view—so forcefully and elegantly described in Gould's *Wonderful Life*—is vigorously disputed in British paleontologist Simon Conway Morris's 1998 book *Crucible of Creation*, about the Burgess Shale and Cambrian Explosion. Conway Morris is, ironically enough, a central and sympathetic figure in Gould's book. He is one of the architects of our new understanding of the Cambrian Explosion. But Conway Morris denies that disparity has been decreasing since the Cambrian, citing several cases that suggest just the opposite. He also attacks Gould's metaphor of "replaying the tape" by showing how convergence in evolution (where distinct lineages evolve similar body types in response to similar environmental conditions) can produce the same types of body plans from quite unrelated evolutionary lineages. Conway Morris argues that even if the ancestor of vertebrates went extinct during or soon after the Cambrian, it is likely that some other lineage would have evolved the body plan with a backbone, because this design is optimal for swimming in water. This view is quite antithetical to that espoused by Gould. We thus have several models for diversification (see Figure 7.3), and how it actually occurred on Earth is still in doubt.

AFTER THE CAMBRIAN EXPLOSION: THE EVOLUTION OF DIVERSITY

Another aspect of the Cambrian (and one that we admit to rather cavalierly omitting) is that not only did diversity and complexity of species increase

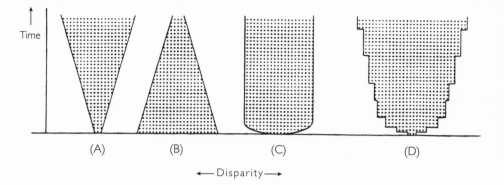

Figure 7.3   *Various interpretations of the history of life and its disparity. (A) The traditional view, whereby disparity steadily increases through geological time. (B) The view presented by S.J. Gould, whereby maximum disparity occurred in the Cambrian. (C) The view that disparity increased very rapidly in the Cambrian and thereafter stayed much the same. (D) The view that disparity increased rapidly in the Cambrian and since then has generally increased, though at varying rates. (From Simon Conway Morris.)*

through time, but the ecosystems in which they lived changed as well. The evolution and emergence of eukaryotic creatures, culminating in the Cambrian Explosion, was accompanied by a shift from bacterial ecosystems to assemblages far more diverse and complex. The dramatic decline of stromatolites, the layered bacterial structures that were so common until a billion years ago, may be evidence of this transformation from a prokaryote-dominated to a eukaryotic world. With the rise of animals, efficient herbivores appeared, and the passive bacterial mats that we call stromatolites when they are fossilized, served as food for the emergent herbivores.

The Cambrian was the time when the most profound of these changes occurred. Yet it was not the last period of major diversification. Paleontologist Jack Sepkoski of the University of Chicago has spent more than two decades analyzing the diversity of organisms through time. He identifies two significant episodes of diversification after the Cambrian: one in the lower Ordovician (the period immediately succeeding the Cambrian) and one at the start of the Cenozoic era, the time interval some 65 million years ago that immediately followed the great extinction that wiped out the dinosaurs and so many other species. The major unanswered question is whether the great rise in species that characterized the last 500 million years was inevitable once animals arose or was itself due to chance.

RELEVANCE TO THE FREQUENCY OF LIFE ON OTHER PLANETS

Is the long wait for animals—the story of animal life on Earth—the exception or the rule for any other planets with emerging life? Could either oxygenation or the evolutionary steps necessary for completion of the complex animal body plan have occurred more rapidly on some other planet, and if so, under what conditions? The lesson of Earth's Cambrian Explosion is that two parallel preparatory steps must be taken if complex metazoans—animals—are to appear. First, an oxygen atmosphere must be constructed. This is surely the most critical environmental step. Second, a very large number of evolutionary adaptations must be concluded to allow the evolution of an ocean liner—our animals—from the toy sailboat—the bacteria—that began it all.

Both of these parallel tracks require time. There do not appear to be any shortcuts. On Earth, one or both required several billion years. And during that time, Earth had to maintain a temperature that allowed the presence of liquid water and avoid what we might call "planetary disasters" of sufficient magnitude to sterilize the evolving root stocks of animals. In the next chapter, we shall see why no such disaster put an end to animal evolution on planet Earth.

Mass Extinctions
and the
Rare Earth
Hypothesis

Much of the work we do as scientists involves filling in the
details about matters that are basically understood already,
or applying standard techniques to new specific cases. But
occasionally there is a question that offers an opportunity
for a really major discovery.

—Walter Alvarez, *T. Rex and the Crater of Doom*

Imagine that we are in a spaceship orbiting Earth 65 million years ago—
about 500 million years after the Cambrian Explosion described in the last
chapter—on the day that an asteroid enters the atmosphere and streaks
down toward what is now the Yucatan region of Mexico. We are about to
witness the collision that will eliminate the dinosaurs (and 60% of all other
species) from the rolls of the living.

The asteroid (perhaps it is a comet) is between 6 and 10 miles in diameter, and it enters the Earth atmosphere traveling at a rate of about 25,000 miles an hour. At such speed the body takes 10 seconds to pass through the atmosphere and then smashes into Earth's crust. Upon impact, its energy creates a non-nuclear explosion at least 10,000 times as strong as the blast that would result from humankind's entire nuclear arsenal detonating simultaneously. The asteroid hits the equatorial region in the shallow sea then covering the Yucatan and creates a crater as large as the state of New Hampshire. Thousands of tons of rock from the ground-zero impact area, as well as the entire mass of the asteroid itself, are blasted upward. Some of the debris goes into Earth orbit, while the heavier material reenters the atmosphere after a suborbital flight and streaks back to Earth as a barrage of meteors. Soon the skies over the entire Earth glow dull red from these flashing small meteors. Millions of them fall back to Earth as blazing fireballs and, in the process, ignite the verdant Late Cretaceous forests; over half of Earth's vegetation burns in the weeks following the impact. A giant fireball also expands upward and laterally from the impact site, carrying with it additional rock material that clots the atmosphere as fine dust is transported globally by stratospheric winds. This enormous quantity of rock and dust begins sifting back to Earth over a period of days to months. Great dust plumes and billowing smoke from burning forests also rise into the atmosphere, cloaking Earth in a pall of darkness. From space we begin to lose sight of the surface of the planet and can see only darkening gauze obscuring Earth's once green and blue surface. It is a vision from Dante's *Inferno*, a nightmare of red fires and black soot.

The impact creates great heat both on land and in the atmosphere. The shock heating of the atmosphere is sufficient to cause atmospheric oxygen and nitrogen to combine into gaseous nitrous oxide; this gas then changes to nitric acid when combined with rain. A prodigious and concentrated acid rain begins to fall on land and sea, and before it ends, the upper 300 feet of the world's oceans are acidic enough to dissolve calcareous shell material. The impact also creates shock waves spreading outward through the rock from the festering hole in the crust; Earth is rung like a bell, and earthquakes of unprecedented magnitude occur. Huge tidal waves spread outward from the impact site, eventually smashing into the continental shorelines of North America, and perhaps Europe and Africa as well, leaving, when they recede, a trail of destruction and a monstrous deposit of beached and bloated dinosaur carcasses skewered on uprooted trees. The surviving scavengers of the world rejoice. The smell of decay is everywhere.

For several months after this fearsome day, no sunlight reaches Earth's surface; the atmosphere is darker than the oil-fueled miasma that blanketed Kuwait following the Gulf War. After the initial rise in temperature from the blast itself, the ensuing darkness causes temperatures to drop precipitously over much of the planet, creating a profound winter in a previously tropical world. The tropical trees and shrubs begin to die; the creatures that live in them or feed on them begin to die; the carnivores that prey on the smaller herbivores begin to die. The Mesozoic era, which began 250 million years after the Cambrian Explosion featured in the last chapter, comes to the end of its nearly 200-million-year reign.

Following months of darkness, Earth's skies finally begin to clear, but the extinction—the death of myriad species—is not yet finished. The impact of winter comes to an end, and global temperatures begin to rise—and rise. The impact has released enormous volumes of water vapor and carbon dioxide into the atmosphere, which now create an intense episode of greenhouse warming. Climate patterns change quickly, unpredictably, and radically around the globe before Earth's temperature regains some equilibrium. From tropical to frigid, then back to even more tropical than before the impact, all in a matter of a few years. The temperature swings produce more death, more extinction.

All of this havoc creates death: the death of individuals, the death of species, the death of entire families of organisms. This event is a planetary catastrophe. Had the impacting object been only twice the size it was, it might have sterilized the surface of planet Earth. It was a narrow escape for complex metazoans.

Just 65 million years ago, such an impact event did end the Mesozoic era, and it ended the Age of Dinosaurs as well. It was but one of many impacts and other assorted global catastrophes that have imperiled complex life on Earth over the past 500 million years. Such events must happen on planets elsewhere in the Universe, and they would surely be the greatest obstacle to the continued existence of any complex metazoan that might exist there. Extinction events are an important aspect of the Rare Earth Hypothesis. Although the animals and plants of Earth have suffered grievously in the assorted mass extinction events through time, the damage could have been worse—and on many other planets where life may have evolved, it probably has been, or will be. If hit at an inopportune time, a planet's higher life might be snuffed out—or it might never be allowed to evolve in the first place.

As we saw in the last chapter, the Earth of 500 million years ago was teeming with complex animals and plants. *Attaining* such a world, for the first time populated by animals, required a large number of evolutionary and environmental changes and took 3 to 3.5 billion years. *Maintaining* these organisms required other conditions. Complex metazoans tolerate a far narrower range of environmental conditions than do microbes; there are no extremophile or anaerobic complex metazoans, for example. Complex metazoans are also far more susceptible to extinction caused by short-term environmental deterioration.

DEFINING MASS EXTINCTIONS

The frequency of animal life in the Universe must be some function of how often it arises and of how long it survives after evolving. We believe that both of these factors are significantly influenced by the frequency and intensity of what are termed mass extinctions, brief intervals when significant proportions of a planet's biota are killed off. There is no mystery about what kills organisms: too much heat or cold; not enough food (or other necessary nutrients); too little (or too much) water, oxygen, or carbon dioxide; excess radiation; incorrect acidity in the environment; environmental toxins; and other organisms. Mass extinctions occur when one or some combination of these factors kills a significant percentage of the planet's biota. There has been no shortage of them in the past.

Mass extinctions have the potential to end life on any planet where it has arisen. On Earth there have been about 15 such episodes during the last 500 million years, 5 of which eliminated more than half of all species then inhabiting our planet. These events significantly affected the evolutionary history of Earth's biota. For example, if the dinosaurs had not suddenly been killed off following a comet collision with Earth 65 million years ago, there probably would not have been an Age of Mammals, because the wholesale evolution of mammalian diversity took place only after the dinosaurs were swept from the scene. While dinosaurs existed, mammals were held in evolutionary check. Mass extinctions are thus both instigators of and impediments to evolution and innovation. Yet much of the research into mass extinctions suggests that their disruptive properties are far more important than their beneficial ones. If planets with life are gardens, then the mass extinctions are the pests and droughts as well as, perhaps, the fertilizer. Yet as any gardener knows, plants are most susceptible when they are young, and disasters are most pronounced early in the growing season. A late frost, a catastrophic hailstorm, the emergence of early spring pests, a lack of sun—all make the early growing season the most hazardous time. So, too, with animal life on any planet. We believe that the early period in the evolutionary history of complex metazoans is by far the most dangerous interval. In our view, planetary disasters (resulting in mass extinctions) that occur *before* the evolution of complex metazoans and those that occur *after* they are established through the process of species diversification are far less likely to end in the extinction of all life. The fossil record of life on Earth supports this prediction in that the Cambrian period, when complex animal life had recently evolved, shows the most significant losses of higher taxa.

Unlike animals, which are fragile and easily killed, microbes are less susceptible to mass extinction events. Once established as a deep microbial biosphere, the bacterial grade of life is probably very difficult to eradicate. Short of planetary sterilization through destruction of the planet by a supernova or collision with a very large asteroid, the deep microbial biosphere of any planet must act as an effective reserve of life, because the regions several kilometers beneath the surface are insulated from even prodigious disasters that affect the surface regions. Surface life, on the other hand (even bacterial surface life), is surely susceptible to major planetary catastrophes, such as the impact of truly large comets or asteroids. It may be that life on Earth's surface was repeatedly sterilized during the period of heavy bombardment about 4 billion years ago, only to be reseeded by the deep-earth microbes or by the return of rocks ejected by the collisions. But for animal life, quite the opposite is true. Animals are not capable of the safer subterranean existence or of hibernating in the vacuum of space. If they are wiped out by catastrophe, they cannot be immediately restocked from some underground reserve. They have to evolve again in a slow, step-by-step process that lasts hundreds of millions or even billions of years.

On every planet, sooner or later, a planetary catastrophe can be expected that either seriously threatens the existence of animal life or wipes it out altogether. Earth is constantly threatened by planetary catastrophe—mainly by impact from comets and asteroids crossing the Earth's orbit, but also from other hazards of space. Yet it is not only the hazards of outer space that threaten the diversity of life on this planet and on any others where it exists. There are Earth-borne causes of catastrophe as well as extraplanetary causes. Both types have brought about mass extinction on this planet in the past and would be likely to do so on other planets as well.

Types of Planetary Disasters

The immediate, or direct, cause of all mass extinctions appears to be changes in the "global atmosphere inventory." Changes to the atmospheric gases (which may be changes in volume or in the relative constituents of the atmosphere) can be caused by many things: asteroid or comet impact, degassing of carbon dioxide or other gases into the oceans and atmosphere during flood basalt extrusion (when great volumes of lava flow out onto Earth's surface), degassing caused by liberation of organic-rich ocean sediments during changes in sea level, and changes in the patterns of ocean circulation. The killing agents arise through changes in the makeup and behavior of the atmosphere or through factors such as temperature and circulation patterns that are dictated by properties of the atmosphere.

Planetary disasters can occur for a great number of reasons. We shall examine a few, in no particular order of importance.

- *Changing a planet's spin rate*

We take Earth's 24-hour spin rate for granted, when in fact it appears to be highly anomalous if we compare it to other planets and satellites in our solar system. Jupiter and Saturn, for instance, each far greater in mass and diameter than Earth, spin much faster. Many other planets, however, such as Venus and Mercury (and even our own Moon), spin much more slowly, such that they always present the same face to the body they revolve around. In lower-mass stars, planets in the habitable zone become "tidally locked" by the gravitational force of the larger star or planet they revolve around. When one side always faces the star in question, that particular face becomes very hot, whereas the other side is always facing cold space and becomes frigid. Either environment would be lethal to surface life and prevent its evolution.

Planets can change spin rates, and when they do, any life already adapted to a particular spin regime would be likely to face planetary disaster because of the major temperature changes it would encounter. Earth itself has been gradually slowing, a phenomenon that has probably altered the distribution of cloud cover over time.

- *Moving out of the animal "habitable zone"*

Animal life needs liquid water, so it requires a mean global temperature that allows liquid water to exist. Any movement of a planet out of an orbit that allows such temperatures will create a planetary disaster. Though such changes of orbit are unlikely, they could be caused by another planet in a stellar system. Such perturbations would be common in open star clusters.

- *Changing the energy output of the sun (star)*

Complex animal life on any planet is dependent on stellar energy. If stellar output either increases or decreases such that liquid water can no longer exist, the result will be disastrous to animal life or to the prospects for its evolution. Short-term and long-term changes in stellar energy output may be one of the most common forms of planetary extinction—and even sterilization. Some scientists are convinced that the end of life on Earth will be caused by an increase in the sun's energy output. This is nothing new. As we have seen, the amount of energy being produced by the sun—and indeed, by most stars—increases over time. On Earth, the maintenance of a relatively constant temperature has been attained through a gradual reduction in greenhouse gases as the amount of energy from the sun has increased, thus keeping temperatures in check. We seem to be nearing the end of this type of planetary temperature regulation, however. There are now very small volumes of carbon dioxide in the atmosphere compared to earlier periods of geological time, and the sun's energy output continues to increase. Some scientists have predicted that temperatures on Earth will become too high for animal life within several hundred million years from now. That event, when it comes to pass, will produce the last greatest mass extinction on Earth, its sterilization.

- *Impact of a comet or asteroid*

Any planetary system is rife with cosmic debris: asteroids and comets, the residue left over from planetary formation. Great quantities of this material will eventually strike all members of a planetary system, and the energy released can spell planetary disaster. Such disasters are now known to have caused mass extinctions on Earth. In 1980, Luis and Walter Alvarez, Frank Asaro, and Helen Michel from the University of California at Berkeley proposed that one of the greatest of all mass extinctions, the 65-million-year-old event that killed off the dinosaurs and many other species living near the end of the Mesozoic Era, was caused by the impact of a large meteor or comet striking Earth, as described at the beginning of this chapter. As evidence for this view mounted, most scientists realized that collision with a meteor or comet could cause a biotic crisis on any planet and that it has done so at least once (and probably other times as well) during Earth's past (see Figure 8.1).

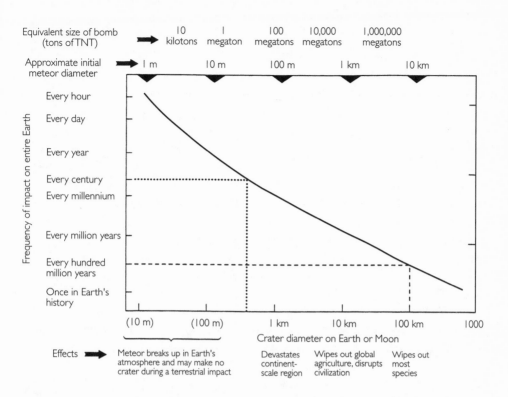

Figure 8.1   *The rate of meteor impacts at the top of Earth's atmosphere as a function of meteor size. Bottom scale gives crater size for typical impact velocity of about 15 kilometers per second. Top scale gives meteorite size and energy in terms of tons of TNT. Dotted line shows the scale of the Siberian meteoritic explosion of 1908. Dashed line shows the scale of the impact 65 million years ago that wiped out dinosaurs and other species. (After Hartmann and Impey, 1994; 1993 data from E. Shoemaker, C. Chapman, D. Morrison, G. Neukum, and others)*

Many variables affect the degree of lethality resulting from a collision, such as the meteor's size, composition, angle of impact, and velocity and the nature of the impact target area. In the case of the Cretaceous event (also known as the K/T impact), for instance, the target rock was rich in sulfur, which exacerbated the impact's environmental effects. (The sulfur reacted with air and water to produce a highly toxic acid rain that lasted many months after the impact event itself.) Moreover, not only the geology of the impact site, but also its geography, may play an important part. An impact in

165

a low-latitude site will have entirely different consequences from a similar body hitting a high-latitude site at a similar angle and speed, because the distribution of lethality across the globe may be produced by atmospheric circulation patterns. Finally, the nature of both the biota and the atmosphere at the time of impact are surely important. An impact in a highly diverse world of ecological "specialists"—animals and plants with little tolerance for environmental change—might produce more extinction than the same event in a low-diversity world of "generalists." And an impact in a greenhouse world might have different effects from one where greenhouse gas inventory or oxygen content was lower than that on Earth today.

In the early years after the Alvarez hypothesis was advanced, some investigators thought that a general synthetic model linking most or all mass extinctions to impact would emerge. This was the thinking behind the "Nemesis" hypothesis of astronomer Rich Muller from Berkeley, and it underlies the work of David Raup and Jack Sepkoski of the University of Chicago, who hypothesized in 1984 that mass extinctions show a 26-million-year periodicity. Since then, elevated levels of iridium (the platinum-group element used by the Alvarez team as a sign of impact) have been found from 11 different time intervals in the geological record. Yet most of these are at such low concentration that they are not indicative of larger impacts. The evidence to date suggests that only the major mass extinction at the end of the Triassic and that at the end of the Cretaceous Period (the K/T) were brought about by the effects of impact.

The presence of numerous impact craters on every stony planet or moon of the solar system is stark evidence of the frequency of these events, at least early in the history of our solar system. It is probable that impact is a hazard in most, or perhaps all, other stellar systems as well. Impacts are probably the most frequent and important of all planetary catastrophes. They could completely reset the course of the biological history of a planet by removing previously dominant groups of organisms, thus opening the way for entirely new groups or for the rise to dominance of previously minor groups.

166

- *Nearby supernova*

Another mechanism that could produce a mass extinction is the occurrence of a supernova in the sun's galactic neighborhood. Two astronomers from the University of Chicago calculated in 1995 that a star going supernova within 10 parsecs (30 light-years) of our sun would release fluxes of energetic electromagnetic and charged cosmic radiation sufficient to destroy Earth's ozone layer in 300 years or less. Much recent research on ozone depletion in the present-day atmosphere suggests that removal of the ozone layer would prove calamitous to the biosphere and to the species residing within. A depleted ozone layer would expose both marine and terrestrial organisms to potentially lethal solar ultraviolet radiation. Photosynthesizing organisms, including phytoplankton and reef communities, would be particularly affected.

Judging by the number of stars within 10 parsecs of the sun in the last 530 million years, and by the rates of supernova explosions among stars, astronomers have concluded that it is *very* plausible that one or more supernova explosions have occurred within 10 parsecs of Earth during the last 500 million years. They also believe such explosions are likely to occur every 200 to 300 million years. The probability of nearby supernovae would be much greater closer to the galactic center, as suggested in Chapter 2.

- *Sources of gamma rays*

Astronomers have detected sudden bursts of intense gamma radiation being emitted from various galaxies (gamma rays are the most dangerous radiation emitted by atomic bombs). Although very little is yet known about these short but extremely violent releases of energy, they would be lethal to any life on nearby planetary systems.

- *Cosmic ray jets and gamma ray explosions*

A new entry into the mass-death rogues gallery is lethal bursts of radiation produced by violent stellar collisions. Cosmic ray jets and gamma rays might both result from the same source: merging neutron stars. Astronomers Arnon Dar, Ari Laor, and Nir Shaviv have postulated that cosmic ray jets may account for several of the major mass extinctions and might explain the rapid evolutionary events that follow them. They propose that high-energy fluxes of cosmic rays follow the merger or collapse of neutron stars, themselves the residues of supernovae. These explosions are the most powerful in the Universe, releasing in a few seconds as much energy as the entire output of a supernova. When two of these objects coalesce, they create a broad beam of high-energy particles that, if it hit Earth, would be capable of stripping away the ozone layer and bombarding the planet with lethal doses of radiation.

The frequency of these events is the critical issue. New calculations suggest that these events may be both more frequent and more dangerous to life in any galaxy than previously supposed. Chicago physicist James Annis proposed in 1999 that gamma ray explosions are so lethal that a single such event could obliterate life over much or all of an entire galaxy. Annis has calculated that the rate of such explosions is about one burst every few hundred million years in each galaxy. For instance, Annis suggests that if the energy from such an event hit Earth, it would kill all land life on our planet, even if the explosion occurred at the center of our galaxy. If such violent and dangerous collisions are rare, they are but one more low-probability event. Yet both Annis and Dar argue that such collisions occur relatively often and were even more common earlier in the history of the Universe. They calculate that such effects would cause a major mass extinction on Earth every hundred million years.

- *Catastrophic climate change: Icehouse and Runaway Greenhouse*

Under certain circumstances, radical changes in climate can cause mass extinction. Major glaciations and greenhouse heating are examples, and both depend on the amount of carbon dioxide or other greenhouse gases in the atmosphere. These are the actual killing mechanisms brought about by the reduction or increase of stellar output or by a planet's orbit becoming either closer to or farther from its sun. Climate changes intense enough to threaten the biosphere with major mass extinction would involve great swings in mean planetary temperature, as well as relocation of oceanic current systems and shifts in planetary rainfall patterns.

The two most catastrophic such conditions can be called Icehouse (the Snowball Earth events are examples) and Runaway Greenhouse. In both

168

cases, global temperatures move outside the 0–100°C range that allows the presence of liquid water on the planet. We will see possible examples of each when we examine the fates of Venus and Mars in the next chapter.

- *The emergence of intelligent organisms*

There is abundant evidence that the emergence of humanity as a globally distributed species armed with technology has triggered a new episode of mass extinction on Earth. It can be argued that the emergence of any intelligent species co-opting a planet's resources in the service of advanced technology and agriculture will necessarily cause a planetary mass extinction.

THE FREQUENCY OF MASS EXTINCTIONS

How often do mass extinctions take place? Perhaps the best way to address this question is to use the same methods that meteorologists and hydrologists use in assessing the risk from weather and floods. Many natural phenomena—such as floods, earthquakes, and droughts—are distributed through time in a similar way. Small events are common, large ones rare. The best way to see how frequent the really rare events are, is to assemble all the available data and arrange them by return time or waiting time. For instance, we might ask how often in a century, or in a thousand years, a flood of some given intensity occurs? We can then define "10-year" floods (events of such magnitude that we can expect one every 10 years on average) and compare them to much larger, 100-year events and the even more catastrophic 1000-year events. This does not mean that we cannot get two 100-year events in successive years, only that the probability of two such events taking place in successive years is vanishingly small. Hydrologists use a technique called extreme-value statistics to extrapolate waiting times beyond the length of historical records. These estimates are, of course, imperfect. But they allow scientists to make estimates about, say, 1000-year events when only 100 years of historical records are available.

Paleontologist David Raup has adapted this same technique to investigating questions of mass extinction. Raup's questions are very similar to those posed by meteorologists interested in estimating how often giant floods may take place. Raup notes that we have a good record for the past 600 million years of Earth history, so we can define the 10-million-year and 30-million-year events with confidence. Using these statistics, Raup calculated what he called a kill curve.

The kill curve is a graph showing the expected waiting times for mass extinctions of varying magnitude. It depicts the average "species kill"—the percentage of all species on Earth at a given time suddenly going extinct as a result of some mass extinction event—for a series of waiting times. Raup's curve is not entirely theoretical; he derived it by first amassing the extinction records of more than 20,000 genera of organisms on the basis of their actual geological longevities. Raup used the *Zoological Record*, a compendium of the life now on Earth, found the first occurrence and last occurrence of all of these genera of organisms, and then tabulated the results in a colossal database. His data are thus derived from our best information about the actual geological ranges of the organisms he surveyed.

The kill curve gives us a sense of how many species go extinct over a given period of time. According to this curve, there is negligible extinction in the case of natural phenomena that occur about every 100,000 years. The million-year event is more consequential, with perhaps 5% to 10% of all species on Earth going extinct. That figure rises to 30% of all species at the 10-million-year event and to nearly 70% of all species at the 100-million-year event. These are frightening numbers. If nearly three-quarters of all species go extinct in a short-term planetary catastrophe of some sort every 100 million years, it suggests that we are living on quite an unsafe planet.

Raup discusses that last concern in the 1990 book *Extinction: Bad Genes or Bad Luck*. How often might we expect the event that kills off the entire biosphere of the planet—the complete sterilization of Earth of all its huge diversity of living things? "I once tried extreme-value statistics on extinction data to ask, 'How often should we expect extinction of all species on Earth?.' I don't have much confidence in the results, but they are at least comforting: Extinctions sufficient to exterminate all life should have an average spacing of well over 2 billion years."

Yet this should not be a comforting figure. Indeed, it goes to the heart of the Rare Earth Hypothesis. If we might expect a planetary catastrophe to exterminate all life on this planet every 2 billion years, and if life has already lasted 4 billion years, we are truly pressing our luck! And luck may be just what animals need to evolve for a long time when the grim reaper of planetary extermination is put off only through blind chance.

THE EFFECT OF MASS EXTINCTIONS

Are mass extinctions that stop short of complete sterilization necessarily detrimental to planetary diversity? Perhaps it can be argued that instead of being deleterious to diversity, they are actually forces that *increase* diversity. For example, it can be argued that the various Paleozoic extinctions caused archaic reef communities to be reassembled with more modern types of corals. Mass extinctions paved the way for a takeover of bottom communities previously dominated by brachiopods (archaic shellfish) by the more modern (and more diverse) mollusks. In another case, the extinction of the dinosaurs paved the way for the evolution of many new types of mammals, and it appears that there are more types of mammal species than there were dinosaur species. If these mass extinctions had not occurred, would planetary diversity (the number of extant species) be higher or lower than it is today, other variables (the history of continental drift, for instance) remaining the same?

We can illustrate the enigma of mass extinctions and their effect on global biodiversity as follows: The Cambrian Explosion results in a sudden rise in diversity, followed by an approximately steady state during the Paleozoic. The mass extinctions during the Ordovician and Devonian cause short-term drops in diversity, but these are soon compensated for by evolution of new forms. The great mass extinction that ended the Permian creates a longer-term deficit in diversity, but eventually, in the Mesozoic era, it also is compensated for. In fact, after every mass extinction that occurred on Earth over the past 500 million years, biodiversity has not just returned to its former value but has exceeded that value. Today, in our world, biodiversity is

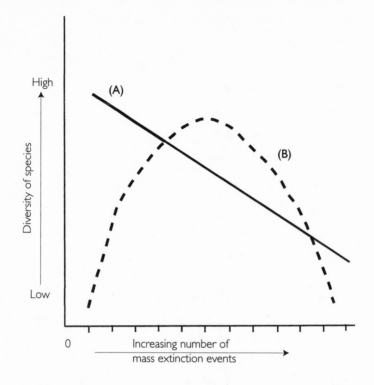

Figure 8.2 *Two models of how mass extinction affects diversity. (A) Increasing number of mass extinction events leads to ever lower diversity. (B) There is some critical number of mass extinction events that causes lower diversity.*

higher than it has been at any time in the past 500 million years. If there had been twice the number of mass extinctions, would there be an even higher level of diversity than there is on Earth now? Perhaps mass extinctions exert a positive effect, creating new opportunities and fostering evolutionary innovation by weeding out decadent or poorly adapted but entrenched and resource-hogging species. On the other hand, perhaps just the opposite is true: If the mass extinctions had not occurred, biodiversity would be higher than it is today (see Figure 8.2). How do we choose?

Interesting as this question is, it has not yet been tested in any way. The fossil record, however, does yield some clues that mass extinctions must be entered on the deleterious rather than the positive side of the biodiversity

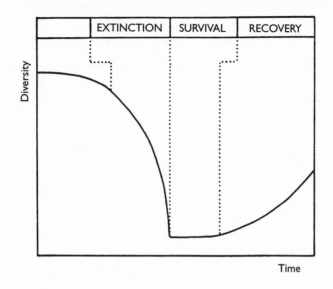

Figure 8.3   *Phases of a mass extinction crisis. (After Kauffman, 1986; based on information in Donovan, 1989.)*

ledger. Perhaps the best clue comes from the comparative history of reef ecosystems following mass extinction. Reefs are the most diverse of all marine habitats; they are the rainforests of the ocean. Because they contain so many organisms with hard skeletons (in contrast to a rainforest, which bears very few creatures with any fossilization potential), we have an excellent record of reefs through time. Reef environments have been severely and adversely affected by all mass extinctions. They suffered a higher proportion of extinction than any other marine ecosystem during each of the six major extinction episodes of the last 500 million years. Reefs disappear from the planet after each mass extinction, and it usually takes tens of millions of years for them to be reestablished. For instance, there were no reefs whatsoever after the Cambrian, Ordovician, Devonian, Permian, Triassic, and Cretaceous mass extinctions. When they do come back, they do so very gradually. It appears that complex ecosystems take a long time to build and to rebuild (see Figure 8.3). When reef systems eventually reappear, they are composed of entirely new suites of creatures. The implication is that mass extinctions, at least for reefs, are highly deleterious and yield net deficits of biodiversity.

RISK AND COMPLEXITY

Can we discern a relationship between the complexity of life and its risk of succumbing to mass extinction? Recent evidence suggests that as organisms become more complex, their risk of undergoing extinction increases. As an organism's complexity increases, so too does its fragility; hence increasing complexity in most cases narrows the environmental tolerances of a living organism. Any bacterium can withstand the rigors of outer space (at least for a short period of time), but no animal can. As we move from bacterial life forms to protozoa and then to metazoans, the range of temperature, food supply, and environmental chemistry in which life can persist becomes more restricted.

This generalization seems to apply not only to individuals of a species but also to the species itself. One of the strongest messages communicated by the fossil record is that extinction rate is a function of complexity. On average, simple animals are far more successful at avoiding extinction and thus persist (in terms of geological time) far longer than complex ones; the simpler the species, the longer its reign on Earth. Many bacterial fossils found in 3-billion-year-old rocks are identical to living forms found commonly on Earth today. Are they the same species? Unless we can compare the DNA content of the ancient form with that of its living analog, we cannot tell. But our best guess is that they may indeed be the same species; they certainly have the same external morphology. Simple bacterial species, once evolved, seem to last a long time, perhaps because of their very simplicity and ability to adapt without resorting to new body forms. Complex metazoans, on the other hand, show far shorter ranges, and even among metazoans the inverse correlation between complexity and evolutionary longevity seems to hold. For example, mammals (the most complex animals on the planet) have average species longevity only slightly greater than a million years, whereas bivalve mollusks, which are far simpler, last an order of magnitude longer.

174

But how can complexity be measured? Perhaps a bacterium really is no simpler than a complex metazoan, and the relationship purportedly observed is due to chance or something other than complexity. It turns out that there are ways of comparing complexity. Determining the length of the genome (and the number of genes it includes) is one such way, and an even easier method of characterizing complexity in a metazoan is by describing the number of different cell types it includes, as initially suggested by paleontologist James Valentine.

Zoologists and physiologists have carried out the differentiation of cell types in animals for years. A bacterium or a paramecium has only a single cell, of course, but with the advent of animal life exhibiting multicellularity, various body cells became specialized. A sponge, among the simplest of multicellular animals, has at least four cell types: one for catching food, one for secreting spicules (primitive supporting structures), one for transporting material around the body, and one that acts as a type of skin cell. Vertebrate animals such as ourselves have many more than this; mammals have in excess of 100 cell types.

Oddly enough, no one has yet tried to correlate complexity as measured by number of cell types with evolutionary longevity. The latter, also known as extinction rate, has been calculated for most groups of animals and plants by paleontologist Jack Sepkoski. Here we have combined these two sets of data and searched for correlation. The results seem to substantiate the view that complexity comes at a price of lower evolutionary longevity. This finding suggests that ever more complex animal or plant species show ever shorter evolutionary ranges, whether on Earth or elsewhere. And it suggests that risk of extinction increases through time (see Figure 8.4).

The History of Mass Extinctions on Earth: Ten Events

Paleontologists have discovered many mass extinction events that occurred since the Cambrian Explosion (that is, in the past 540 million years). Yet other mass extinction events of earlier times are largely unknown to us, because they occurred when organisms rarely made skeletal hard parts and thus rarely became fossils. Perhaps the long period of Earth history prior to the advent of skeletons was punctuated by enormous global catastrophes that decimated the biota of our planet—mass extinctions without record. Yet very little attention has been paid to earlier extinction events. For instance, astrobiologist James Kasting believes that the greatest mass extinction of all time may have been brought about by the "Snowball Earth" events of around 750 million years.

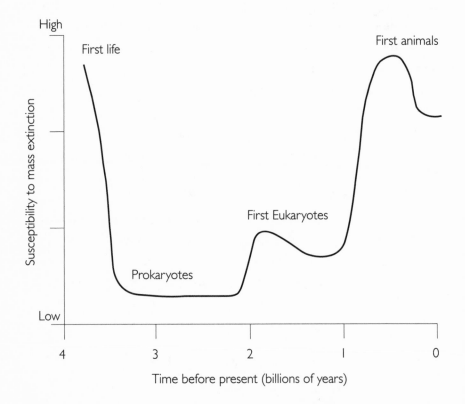

Figure 8.4   *Hypothesized curve of extinction "risk," or susceptibility, through time. Extinction risk is highest soon after a new evolutionary type appears and then lessens as diversification occurs. Diversification is insurance against extinction.*

Of the various mass extinctions known from the last 500 million years (15 events over that time interval are officially classed as "mass extinctions"), six have been especially catastrophic as measured either by the number of families, genera, or species going extinct or by the effect on subsequent biotic evolution. To this list we propose adding three more that occurred before 500 million years ago, as well as the current biodiversity crisis caused by the effects of a runaway human population. This latter mass extinction event is presently under way, so its final extinction total cannot yet be tallied. Nevertheless, it may be representative of what happens whenever an intelligent species arises on a planet with complex metazoans.

Our understanding of the various events is in inverse proportion to their age: The older they are, the more mystery still surrounds them. The modern extinction that is still under way will also be dealt with only briefly. The most recent of the ancient events (the K/T event) is by far the most studied and best known. Accordingly, we will discuss it in greatest detail.

- *The bombardment extinctions, 4.6 to 3.8 billion years ago*

    The period of heavy bombardment is thought to have sterilized Earth's surface at least several times. There is no other information known.

- *The advent of oxygen—Snowball Earth, 2.5 to 2.2 billion years ago*

    The rise of oxygen certainly doomed to extinction most anaerobic bacterial species then on Earth. There is little or no fossil record of this phenomenon, which may have coincided with the first "Snowball Earth" event.

- *Snowball Earth events of 750 to 600 million years ago*

    We have almost no information about these events, which may have included three or four separate extinctions coinciding with repeated glaciations. There did appear to be wholesale extinction among stromatolites and planktonic organisms called acritarchs. The lack of fossilized animals from this time period obscures these events.

- *The Cambrian mass extinctions, 560 to 500 million years ago*

    The extinctions that took place immediately before and then during the Cambrian period remain the most enigmatic of all extinction episodes. We believe that they are also the most important, in terms of their effects on animal life on this planet.

As we saw in an earlier chapter, the "Cambrian Explosion" was the single most significant event in the history of animal life on Earth. In a relatively short interval of time, all of the animal phyla still on Earth appeared. Since the end of the Cambrian, no new phyla have evolved. Yet however celebrated it is for its period of diversification, the Cambrian was also a time of significant extinction. It seems that some of the phyla that appeared in the Cambrian did not survive for long. The Burgess Shale fauna described by paleontologist Stephen Jay Gould contains numerous organisms that seem to belong to no living phylum, and many paleontologists believe that many phyla went extinct in the interval between 540 and perhaps 500 million years ago.

Some scientists argue that the Cambrian Explosion was preceded by the first of all mass extinctions, which caused the disappearance of the Ediacaran fauna. This fauna consisted of the odd assemblage of jellyfish and sea anemone-like animals found in strata immediately below the base of the Cambrian from many parts of the world. They appear to be the first diverse assemblage of animals—perhaps early forerunners of familiar animals such as the Cnidaria and various worms, or perhaps an assemblage of phyla now all extinct. In any case, they disappeared in sudden and dramatic fashion immediately before the Cambrian period. The disappearance of the Ediacaran animals remains a mystery. It may have been brought about through competition with the newly evolving and more modern groups of animals that typify the Cambrian fauna, or they may have been driven into extinction by sudden environmental change.

A second wave of extinctions occurred about 20 million years after the Ediacaran crisis. This second crisis unfolded over several millions of years, and it gravely affected the first reef-forming organisms (called archeocyathids), as well as many groups of trilobites and early mollusks. Again, there is little direct evidence to suggest what may have brought about these extinctions, other than that they seem to be linked to changes in worldwide sea level and to the formation of anoxic bottom water.

The Cambrian extinctions remain an enigma (see Chapter 7). Here we have consequential events without overt cause. How consequential they were is discussed later in this chapter.

- *The Ordovician and Devonian mass extinctions, 440 and 370 million years ago*

During the Paleozoic era there were two other major mass extinctions. About 370 million years ago, in the Devonian period, and more than 430 million years ago, in the Ordovician, major mass extinction events decimated the marine faunas of the time. Because our record of land life is poor for both of these intervals, we have much to learn about the severity of these events on land. However, it is clear that the majority of species in the sea went extinct. Both extinctions eliminated more than 20% of marine *families*.

The causes of both of these extinctions remain obscure. Impact has been proposed as the cause of the Devonian mass extinction, but in spite of intensive searching, little evidence for such a cause has emerged. No evidence of impact exists for the Ordovician extinction either. Anoxia, temperature change, and change in sea level are the favored causes, but it is hard to account for such extensive extinction on this basis alone. These are two major extinction events in search of a cause.

- *The Permo-Triassic event, 250 million years ago*

On the basis of various measures of extinction (percent of existing species, genera, or families eliminated worldwide during the event), the Permo-Triassic mass extinction of 250 million years ago appears to have been the most catastrophic of all mass extinctions that have occurred on Earth. Specialists in compiling extinction records through time, such as Jack Sepkoski and David Raup of the University of Chicago, point out that this particular event stands alone in severity compared to all other such events. More than 50% of marine families died out, and this figure is more than twice that for any other extinction. Estimates of the percent of *species* (belonging to various families) that went extinct in this event vary from nearly 80% to more than 90%. It is clear that the vast majority of animal and plant life on the Earth was extinguished.

Intensive research into the cause of this extinction over the past decade has yielded a clearer view of what has been a perplexing problem: Although it has long been recognized as the most catastrophic mass extinction, its cause has been unknown. It now seems that although several causes contributed to the event, the most important derived from a short-term degassing of carbon dioxide from sediments sequestered on the ocean floor and from volcanic gas emitted during unusually severe volcanic eruptions occurring about 250 million years ago. A sudden release of huge volumes of carbon dioxide directly killed marine organisms by carbon dioxide poisoning and indirectly decimated terrestrial life via sudden and intense global heating. The excess carbon dioxide released into the atmosphere greatly enhanced the greenhouse effect, thus increasing the amount of heat trapped in the atmosphere. A heat spike of perhaps 5 to 10 degrees and from 10,000 to 100,000 years in duration probably caused the terrestrial extinctions.

- *The end-Triassic mass extinction, 202 million years ago*

The end of the Triassic period witnessed a significant mass extinction in which about 50% of genera were eliminated. There is still a very poor record for the fate of land life at this extinction boundary, but it is clear that marine life at this time was extensively and catastrophically affected. This mass extinction, like the K/T event described at the beginning of this chapter, was thought to have been brought about by the impact of a large extraterrestrial body, either comet or asteroid. The Manicouagan Crater in Quebec is 100 kilometers in diameter (compared to the approximately 200-kilometer diameter of the Chicxulub Crater associated with the K/T event). This crater has been dated at 214 million years in age, which is older than the age of the Triassic/Jurassic boundary. Environmental changes other than impact have been associated with this extinction event as well, most notably oceanographic changes creating anoxia in many shallow-water environments at the end of the Triassic. It is difficult, however, to see how such changes could have affected land life, which also suffered significant extinction at this time. The probable cause remains unknown.

- *The Cretaceous/Tertiary boundary event, 65 million years ago*

The mass extinction of the dinosaurs, as well as 50% or more of the other species then on Earth, has been recognized for more than a century and a half as one of the most devastating periods of mass death in Earth's long history. Although numerous explanations for this event have been proposed, an asteroid impact cause is now largely accepted.

In 1969, Dr. Digby McLaren of the Geological Survey of Canada proposed that a largely unrecognized mass extinction had exterminated much sea life about 400 million years ago. McLaren suggested that this extinction was brought about by environmental consequences following the collision of a large meteor with Earth. At that time, McLaren had no actual proof of the impact, such as a suspicious-looking crater of the correct age. Nobel Prize-winning chemist Harold Urey of the University of Chicago also proposed that the great mass extinction that ended the Mesozoic era resulted from a comet, the size of Halley's, colliding with Earth. This proposal too was disregarded for lack of evidence. Soon afterward, however, the Alvarez team at Berkeley repeated the claim, this time providing a wealth of scientific evidence.

The Alvarez hypotheses of 1980 that (1) Earth was struck some 65 million years ago by an Earth orbit–crossing object of some sort (asteroid or comet), and (2) the environmental effects stemming from that impact brought about a mass extinction are today widely (though not universally) accepted. What is unequivocal, however, is that these groundbreaking hypotheses changed the landscape of paleontological research and are of central importance to astrobiology as well. The central paradigm shifts were that asteroid and comet impacts can *cause* mass extinctions, and that mass extinction, through whatever cause, *can occur relatively rapidly.*

Like most great science, the Alvarez theory is conceptually simple. Judging by the concentrations of platinum-group elements found in three Cretaceous/Tertiary boundary layers (or K/T boundary layers, as they came to be called) from Italy, Denmark, and New Zealand, the Berkeley group proposed that an asteroid at least 6 miles in diameter struck Earth 65 million years ago and that the environmental aftereffects of this impact caused the great mass extinction. The ultimate killer, according to the Berkeley group, was a several-month period of darkness, or blackout, following the impact. The blackout was due to the great quantities of meteoric and Earth material thrown into the atmosphere after the blast, and it lasted long enough to kill off much of the plant life then living on Earth, including the plankton. With the death of the plants, disaster and starvation rippled upward through the food chains.

By 1984, high iridium concentrations had been detected at over 50 Cretaceous/Tertiary boundary sites around the world. Shocked quartz grains were also discovered from many Cretaceous/Tertiary boundary sites. Fine particles of soot were found disseminated in the same K/T boundary clays. The type of soot found comes only from burning vegetation, and the quantity of soot ultimately found in boundary clays from many parts of the globe suggested that at some time about 65 million years ago, much of Earth's surface was consumed by forest and brush fires. A truly horrifying vision emerged: Soon after the impact, the majority of plants then on Earth apparently burned. Some of the fires may have started from the fireball and searing heat produced by the impact, but the majority were probably set days later as rocky fragments, initially blown into orbit by the explosive force of the titanic impact, streaked back to Earth as bright missiles of destruction.

There was but a single comet or asteroid strike. The impact created a crater between 180 and 300 kilometers wide and now named Chicxulub. The impact target geology (especially the presence of sulfur-rich sedimentary rocks in the target area) may have maximized subsequent killing mechanisms. Worldwide changes in the atmospheric gas inventory, temperature drop, acid rain, and global wildfires are all proposed killing mechanisms.

The litany of disasters related above is enough to make us wonder how any complex metazoans survived this event. Given an impact object of sufficient size, it is clear that impacts such as that which caused the Cretaceous/Tertiary mass extinction are capable of killing off all complex metazoans on a planet. If the object that caused the K/T extinction had been only twice the size it was, that might have been Earth's fate.

- *The modern extinction*

Suggesting that we have entered a mass extinction episode would have been controversial at best during the 1980s, but most investigators now concede that the number of extinctions that have occurred since the end of the last glacial period, some 12,000 years ago, clearly defines the Holocene time interval as one of pronounced and elevated extinction rates. There are many estimates of how many species are currently going extinct each year, though there are few hard data for many regions. What is clear is that the world's forests are being felled inexorably to make way for agriculture and that the removal of forests leads to extinction. At the moment the seas are more insulated, and there is little evidence of major extinction occurring there at present, although this could change quickly as pressure on the world's fish stocks increases. As runoff and chemical pollution increase over the next several centuries, the extinction rate in the sea may rise substantially. Estimates of the faunal tally vary, but all carry the grave message that Earth is losing a great number of species rather quickly. Perhaps the most sobering estimate comes from Peter Raven of the National Academy of Sciences, who has suggested that two-thirds of the world's species may be lost by the year 2300.

The ultimate cause of this extinction is the runaway population of *Homo sapiens*.

COMPARING THE SEVERITY OF THE MASS EXTINCTIONS

The conventional means of comparing the severity of the various mass extinction events has been to compute the percentage of taxonomic categories that went extinct. This monumental work has been carried out largely by paleontologists at the University of Chicago, through literature research initiated by

David Raup and Jack Sepkoski. The first tabulations were of families of marine animals. Several years later Sepkoski tabulated the number of genera going extinct, and he is now working on the number of species. It was through the use of such statistics that the "Big Five" (the Ordovician, Devonian, Permian, Triassic, and Cretaceous extinctions) were differentiated from other Phanerozoic extinctions. If the number of genera going extinct is used as a means of comparing the mass extinctions, then the Permo-Triassic event was the most catastrophic, followed by the Ordovician, Devonian, Triassic, and end Cretaceous. The Cambrian extinctions do not appear to be so "major." Sepkoski's most recent compilation (which he kindly sent to us as a personal communication in 1997) yields per-family extinction rates of 54% for the Permian, 25% for the Ordovician, 23% for the Triassic, 19% for the Devonian, and 17% for the famous K/T extinction. In two other analyses, however, performed separately by paleontologists Helen Tappan and Norman Newell, the Cambrian extinctions of marine families exceed those of the Permian. The Cambrian extinctions are the most consequential of all; about 60% of marine families went extinct in the Cambrian compared to about 55% in the Permian.

These results pose a dilemma. In the Permian period, the cause of the great mass extinction is quite clear. The continents coalesced into one large supercontinent, greatly affecting worldwide climate and temperature in the process, and near the end of the period an additional, sudden, and catastrophic event occurred as well: Enormous quantities of carbon dioxide were released into the seas and atmosphere, causing a sudden and deadly rise in global temperature. In the Cambrian, however, we have no overt cause. Our best guess is that the garden analogy holds true here. The garden of animals during Cambrian time had only just emerged, and although there were many different types or body plans (more so than now, in fact), there were very few species in each category. Even slight changes in environmental conditions were sufficient to wipe out entire categories, entire phyla. The Cambrian was the riskiest period of all time for animal life. Its extinctions, in our view, were the most important in the history of life on Earth, so the Cambrian should be judged as more important than any other such events, including the Permo-

184

Triassic extinction. There is more to comparing extinctions than simply adding up the number of species killed off.

EXTINCTION RISK THROUGH TIME

Do the risks of extinction vary through time? This question involves two variables: (1) Do the environmental conditions that affect life on a given planet change through time? (2) Does the susceptibility of life to extinction change as its evolution progresses? Just as we have had to modify the concept of the "habitable zone," this question cannot be answered without some qualification related to degree of complexity, for the extinction rates of microbes are far different from those of more complex forms. If, as we surmise, extinction risk or rate varies with an organism's complexity, we might expect rather low extinctions during the long period prior to the evolution of animal life, followed by increasing extinction rates (the percentage of the total biodiversity or species richness going extinct during any interval of time) as animal life evolves. Yet, as we have said, this question involves two variables, the second being the possibility that the frequency or intensity of mass extinctions may change as well. There is much evidence that the proximal causes of extinction—the planetary catastrophes that cause the mass extinctions in the first place—vary through time as well.

The history of major mass extinctions on Earth suggests that only two causes have operated in Earth history: impact and global climate change. There are other phenomena that may also have caused extinctions, such as nearby supernovae, but we have no credible evidence that this latter mechanism has indeed occurred. With regard to impact, there is good evidence that the frequency of impact has changed through time, as we saw earlier in this chapter. The most obvious of these changes was the cessation of major impacts during the period of "heavy bombardment" that lasted from approximately 4.3 to 3.8 billion years ago. But even after this rain of major comets ceased, there is evidence of a long, slow decline in impact rates, as documented by Richard Grieve and others. This decrease would have reduced the overall extinction rate during the same period in which the vulnerability of emerging animal groups was increasing. One could argue that even in the last 500 million years, the time of complex animals, there should have been enough comet or asteroid strikes to exterminate animal life on this planet. That has obviously not happened.

THE TRADE-OFF OF COMPLEXITY: RISK AND DIVERSITY

With complexity can come diversity. Although organisms of greater complexity are more susceptible to extinction, their defense seems to lie in numbers. We are just beginning to study their forms in detail, but it does appear that the number of bacterial species is far lower than the number of insect species. Yet if complexity brings with it the price of greater vulnerability to extinction, how have complex animals and plants survived as long as they have on this planet through the various mass extinctions? There must be some aspect of complex animals and plants that helps protect them from extinction—not as species, but as higher taxa—categories above the species level.

The history of animal phyla during and after the Cambrian Explosion described in Chapter 7 is an example of this. Some paleontologists have suggested that as many as 100 animal phyla may have evolved during the Cambrian period (although the consensus seems to be far fewer than this). Some of these phyla went extinct during the Cambrian or at its end. *Since that time, not a single phylum has gone extinct.* It is probably not a simple case of weeding out the bad from the good, where the survivors were those body plans best suited for our world. Rather, it appears that the surviving phyla have endured subsequent planetary disasters by having large numbers of species. As long as a single species survives, the phylum survives and is in a position to rediversify. In the Cambrian, on the other hand, *all* phyla contained just a few species each; the Cambrian disasters eliminated whole phyla because there was such low species-level diversity within the various phyla. As far as animals are concerned, the Cambrian (or just before) was the most dangerous period in the long history of Earth. Since that time, millions of species have evolved in the various phyla, making them far more "extinction-proof." Diversity—the stocking of body plans through the evolution of numerous species—may be the best protection against extinction.

How do higher animals and plants create and then maintain this diversity? First, they have evolved rapid (compared to bacteria) speciation rates. Because higher organisms use sexual reproduction as the dominant means of reproducing, they create a great deal of variability within populations, upon which natural selection can act. Speciation (the creation of new species) occurs when small populations split off from larger parent populations and can no longer exchange genes. Gradually, adaptation of the smaller population makes it sufficiently distinct from the parent population to prevent successful interbreeding if the two populations come in contact again.

This process is critical to maintaining diversity. Constant formation of new species is needed if diversity is to persist, for there is a relatively high rate of species extinction in all groups of animals and higher plants. Since the Cambrian Explosion, the engine driving the creation of new species has been causing the diversity of complex life on Earth to increase, though long-term gains have been periodically—and temporarily—reversed by the various mass extinctions. During the mass extinction events, it is diversity that saves higher taxa. If a given phylum or major body plan is sufficiently diversified— is represented in many forms living in many different environments—it has a high likelihood of withstanding the extinction event.

A PLANET ON THE BRINK

The "close calls" of planetary catastrophe are repeatedly written large in the rock record of Earth. We have teetered on the brink many times. In this chapter we have recounted the obvious near misses, as manifested by the various mass extinctions. Yet a more subtle if no less sobering record can be found in the changes in atmospheric composition through time.

It is clear that the levels of carbon dioxide and oxygen have changed dramatically during the Phanerozoic (the last 530 million years), and these changes may themselves have been trivial compared to the longer but less readily sampled Precambrian interval of time. The ecological effects of these changes are very poorly understood. During the Paleozoic Era, CO_2 values 20 times that of the present day are now confidently inferred to have been present in the lower Paleozoic, followed by a rapid decline in the Permo-Carboniferous time. The world then underwent a massive glaciation, as its greenhouse conditions gave way to far cooler climates.

Changes in the amount of oxygen in the atmosphere have also been profound, but they are far less well documented (or understood) than for carbon dioxide. Estimating ancient oxygen levels is fraught with uncertainty: One scientific group extracts oxygen from ancient amber, but others decry this method as yielding totally erroneous results. Without direct readings, we must estimate the ancient oxygen values in the atmosphere by studying the rate of organic carbon burial in sediment through time or by studying rates of rock weathering. Neither approach is very satisfactory. Yet if these methods have *any* validity, they tell us that the amount of oxygen has varied over the last 500 million years. For instance, there appear to have been far higher levels of oxygen in the atmosphere around 400 to 300 million years ago; levels as high as 35% by volume (compared to 21% today) are possible. This much oxygen would have made forest fires far more common and more devastating. There were also periods of depressed oxygen, and it is not too farfetched to envision scenarios where lowered oxygen levels had large-scale effects, perhaps even inhibiting the evolution or development of certain forms, such as those with very high metabolisms requiring abundant oxygen.

Thus even on a planet such as ours, the most important of all systems for life—the atmosphere—can be unstable over an interval 100 million years long. Planetary atmospheres can change enough to cause mass extinctions, and maintaining an atmosphere conductive to animal life for the staggering periods of time necessary for animal life to evolve and diversify may be the most difficult feat of all.

A MODEL OF PLANETARY EXTINCTION

We can summarize the implications of Earth's history of mass extinctions with regard to the Rare Earth Hypothesis as follows. Mass extinctions probably occurred rarely during the long period in Earth history when life was only of a bacterial grade. With the evolution of more complex creatures, such as eukaryotic cells, however, susceptibility to extinction increased. With the advent of abundant complex animals in the Cambrian, vulnerability to mass extinction may have reached a peak, because diversity was very low. As more and more species evolved within the various body plans, susceptibility to extinction decreased again.

On any planet, the number of mass extinctions may be one of the most important determinants of where animal life arises and, if so, how long it lasts. In planetary systems with large amounts of space debris—and thus a high impact record—the chance that animal life will arise and persist will surely be much lower than in systems where impacts are few. In similar fashion, inhabiting a cosmic neighborhood where large amounts of celestial collisions, supernovae, gamma ray bursts, or other cosmic catastrophes occur will also reduce a planet's likelihood of attaining and maintaining animal life.

It appears that the best "life insurance" is diversity. In the next chapter we will document our view that plate tectonics, also known as continental drift, has been the main process promoting high diversity among animals on Earth.

The
Surprising
Importance
of Plate Tectonics

I magine that we have a spacecraft capable of swiftly taking us to each planet in the solar system. Our goal on this voyage is to try to determine what features of Earth are essential to animal life. On our voyage, therefore, we are looking for some clue to why animal life has been able to survive on Earth for a time period approaching a billion years. What are those factors that have fostered diversity on Earth?

We begin our celestial survey with Mercury, a cratered world of great heat on the sunlit side and great cold on its dark side of the slowly spinning planet. Yet we quickly find that Mercury is not only free of atmosphere, liquid water, and life but is also volcanically dead. Its surface shows mainly numberless craters of a meteor-ravaged world, scars left by the bombardment by comets and asteroids. In contrast to Earth, in the 4 billion years since that time of stony rain, little of geological importance has happened on this planet. Mercury looks like our Moon.

Next we travel to cloud-covered Venus. Its surface looks curiously young, like the face of a child, yet Venus is the same age as Earth. We find that the crust of Venus appears to have been geologically "resurfaced" in some sort of cataclysmic event that caused its surface to melt sometime in the last billion years. Because of this, the numerous craters we saw on Mercury are less common. But Venus has two other prominent geological features: crustal plateaus and volcanic rises that look something like a string of volcanoes whose cones have been lopped off. There is no animal or plant life and no oceans or liquid water of any kind. The surface of Venus is simply too hot—hot enough to melt solder.

Mars is the next leg of this long voyage, and as we orbit the red and ochre planet, we see astounding volcanoes rising high above a cratered and rock-strewn landscape. These volcanoes (the largest in the solar system) are enormous by Earth standards. But they are relatively few in number—solitary, lonely sentinels dispersed across the planet's face. Curiously, there are no other mountains, no equivalents of the Alps, or even the Appalachians. And there are no seas, no lakes, no rivers, and no liquid water, although many geomorphic features of the planet's surface indicate that water was present here long ago.

With Mars we have finished our survey of the so-called "terrestrial" planets. We have already learned much: No other planet has linear mountain chains. Now we travel toward the outer regions of the solar system, arriving in the realm of the gas-giant planets. First we pass Jupiter, with its writhing, multicolored atmosphere and the distinctive Great Red Spot racing around this rapidly rotating colossus of a planet. There are no land features, for there is no distinct planetary surface, no place where the atmosphere ends and land begins. Jupiter is unsuitable for animal life (as we know it, anyway) because it has no solid planetary surface. Perhaps bacteria-like organisms live within its roiling atmosphere, perhaps not. Its satellites, however, might be places where life has arisen and survived, so we swing by each of the four large "Galilean" moons (so named because they were first seen by the great Italian astronomer Galileo): Europa, Callisto, Ganymede, and Io.

Each is somewhat smaller than Earth, and all have frozen surfaces (although Io has active volcanoes). There are no animals or even liquid oceans here, although the frozen ocean of Europa seems a tantalizing possibility for life simply because liquid water may lie deep below its ice-covered surface. Ganymede and Callisto are also likely to harbor subterranean regions of liquid water or brine.

From Jupiter we travel on to the other gas giants of the solar system: Saturn, Uranus, then Neptune. Like Jupiter, each is a great gas ball without any definable surface, but each has smaller, rocky satellites orbiting it, some cratered, some ice-covered. None has animal life, although Saturn's moon Titan does provide an exotic environment with frigid hydrocarbon liquids at its surface and liquid water at warmer depths.

We finally arrive at Pluto, a solid world, but a world without mountains or volcanoes. Like Mercury (the innermost planet of the solar system), frigid, distant Pluto is devoid of volcanic activity.

As we return to Earth from this trip, we ponder what is unique about Earth that may offer us clues to why animal life exists here but not on other planets and their moons in our solar system. A crucial difference, its seems, is Earth's unique possession of liquid water at its surface. Water, the universal solvent, seems indispensable for animal life. Earth has other unique attributes, too, including its oxygen-rich atmosphere and a temperature range that allows liquid water to exist.

Another unique attribute of Earth at first glance seems extraneous to animal life but may indeed be crucial to it: linear mountain ranges. There are, of course, giant mountains elsewhere in the solar system, the tallest being the great volcano Olympus Mons on Mars. Yet such mountains are always single and never occur in chains, unlike most mountains on Earth. There is no equivalent to the Rockies, the Andes, the Himalayas, or the score of other linear mountain chains we are so familiar with. Even at this crude level of observation, oceans, mountain chains, and life make Earth unique in this solar system. Life has had little to do with creating oceans and mountain chains. Yet these features of Earth may have been crucial to the origin of life. In this chapter, we argue that all three of these precious attributes of Earth are connected in a complex interrelationship. All three, furthermore, may be the result of plate tectonics. This process, the movement of the planetary crust across the surface of the planet, is found in our solar system only on Earth, and it may be vanishingly rare in the Universe as a whole. It is not the mountains as such that are so important to life on Earth but the process that creates them: plate tectonics.

It may seem odd to think that plate tectonics could be not only the cause of mountain chains and ocean basins but also, and most enigmatically, a key to the evolution and preservation of complex metazoans on Earth. But there are several reasons to consider this view. First, plate tectonics promotes high levels of global biodiversity. In the last chapter, we suggested that the major defense against mass extinctions is high biodiversity. Here we argue that the factor on Earth that is most critical to maintaining diversity through time is plate tectonics. Second, plate tectonics provides our planet's global thermostat by recycling chemicals crucial to keeping the volume of carbon dioxide in our atmosphere relatively uniform, and thus it has been the single most important mechanism enabling liquid water to remain on Earth's surface for more than 4 billion years. Third, plate tectonics is the dominant force that causes changes in sea level, which, it turns out, are vital to the formation of minerals that keep the level of global carbon dioxide (and hence global temperature) in check. Fourth, plate tectonics created the continents on planet Earth. Without plate tectonics, Earth might look much as it did during the first billion and a half years of its existence: a watery world, with only isolated volcanic islands dotting its surface. Or it might look even more inimical to life; without continents, we might by now have lost the most important ingredient for life, water, and in so doing come to resemble Venus. Finally, plate tectonics makes possible one of Earth's most potent defense systems: its magnetic field. Without our magnetic field, Earth and its cargo of life would be bombarded by a potentially lethal influx of cosmic radiation, and solar wind "sputtering" (in which particles from the sun hit the upper atmosphere with high energy) might slowly eat away at the atmosphere, as it has on Mars.

What Is Plate Tectonics?

Geologists of the eighteenth and nineteenth centuries had little difficulty understanding the origin of volcanoes: Hot magma from deep within the planet rose to the surface regions and spewed forth lava, ash, and pumice to form a cone. Understanding how *nonvolcanic* mountains and mountain ranges could form, however, was more problematic. Countless hypotheses were proposed. These included buckling of the crust as a result of sediment loading (where the weight of slowly accumulating sediment finally causes the crust to crack in linear fashion), shrinking of the planet (causing ridges to form as on a dried prune), and an expanding Earth (where the expansion creates mountain ranges). In 1910 American geologist Frank B. Taylor proposed a radically new idea: The drifting of continents caused the great mountain chains. This heresy was immediately decried by nearly all other geologists and geophysicists, who could envision no mechanism by which such "drift" could occur.

Taylor's hypothesis, however, kindled a spark of interest that would not die. Soon other scientists began toying with the idea and searching for supporting evidence. The most dogged of the new converts was a German meteorologist named Alfred Wegener, who from 1912 until his death in 1930 on Arctic ice was obsessed with the idea. Drawing on evidence from geology and geophysics, Wegener was the first to show how the fit of various coastlines supported the idea that all the continents were once united in a single "supercontinent." He was also the first to use paleontological evidence to support this claim: He argued that the presence of similar fossil species on land masses now widely separated could have come about only if the various continents had once been in contact. He convinced some other geologists that continents did and do drift, although the majority remained skeptical until the 1960s.

The greatest obstacle to the idea (and the rallying point of all "antidrifters") was the seeming absence of any sort of reasonable, underlying mechanism. How could the massive continents "float" over the surface of the planet's stony surface? The answer to this question, it was eventually discovered, lies in the different phase states of Earth's uppermost layers, known as the crust and upper mantle, and the presence of thermal convection in these regions. Scottish geologist Arthur Holmes first proposed that the upper mantle acts like boiling water, producing large moving "cells" of material. Deep below the surface, the fluid, hot material composing the upper mantle is heated and begins to rise; as it rises it cools, and eventually it begins to flow parallel to the planet's surface. When it cools sufficiently, it sinks again. Holmes proposed that where it rises, the convection cells might rupture the rigid, solid crust and then carry it along, piggyback fashion, in those regions where the mantle moves parallel to the surface.

The outlandish ideas behind the early theory of continental drift were eventually shown to be correct. Evidence came from many sources, including paleontological data and even the fit of the continental coastlines, as first proposed by Wegener. Yet the two most powerful lines of evidence for plate tectonics (another term for continental drift) came from fields unknown to Wegener: From the study of paleomagnetics, which allowed the reconstruction of ancient continental positions, and from oceanographic studies of the ocean floor, which revealed the presence of enormous underwater volcanic centers, areas where the sea floor literally pulls away from itself.

We know now that all continents are masses of relatively low-density rock embedded in a ground mass of more dense material. The low-density rocks have the average composition of granites, whereas the higher-density rocks that make up the ocean crust are basaltic in composition. Because granite is less dense than basalt, the granite-rich continents essentially "float" on a thin (relative to Earth's diameter) bed of basalt. Earth scientists like to use the analogy of an onion; the thin, dry, and brittle onion skin is the crust, sitting atop a concentric globe of higher-density, wetter material. Continents are like thin smudges of slightly different material embedded in the onion skin. Unlike an onion, however, Earth has a radioactive interior and constantly generates great quantities of heat as the radioactive elements, entombed deep within in the planet, break down into their various isotopic by-products. As this heat rises toward the surface, it creates gigantic convection cells of hot, liquid rock in the mantle, just as Arthur Holmes envisioned. Like boiling water, the viscous upper mantle rises, moves parallel to the surface for great distances (all the while losing heat), and then, much cooled, settles back down into the depths. These gigantic convection cells carry the thin, brittle outer layer—known as plates—along with them. Sometimes this outermost layer of crust is composed only of ocean bed; sometimes, however, one or more continents or smaller land masses are trapped in the moving outer skin.

Under the pressures and temperatures encountered at depths many kilometers beneath Earth's surface, the familiar rocks of our crust act in ways very different from what we are used to. Victor Kress of the University of Washington pointed out that all but a tiny fraction of the upper mantle is entirely solid. Yet it acts like a liquid in certain ways, most significantly in its "convection": the process whereby a liquid, when heated, flows upward and then across the top of its container. The mantle convects in the manner of a liquid only because the movement is so slow, and the temperatures so high, that individual crystals have time to deform in response to stress. The upper mantle is a hot, highly compressed mass of crystal that acts like a very viscous liquid.

The "plates" of plate tectonics are composed of all of the crust and a thin section of mantle that underlies it, which together act as a relatively rigid composite layer. Plates are of varying thickness, and their "bottoms" are thought by many scientists to coincide with the 1400°C isotherm (a region where the rock is heated to that very high temperature at which mantle rock material melts into a plastic-like medium). Another way of visualizing the plate foundation is to recognize that this region is characterized by much decreased viscosity. The difference in viscosity between the overlying plate and the underlying region of lowered viscosity is highly important in plate tectonics. It allows the relatively rigid crust to slip as a unit over the zone of high viscosity. Plates composed of oceanic crust and mantle are about 50–60 kilometers thick, whereas the plates with continental crust average about 100 kilometers in thickness.

Let's begin our examination of the plate tectonic process with ocean basins. The crust we find lining the bottom of the world's oceans is largely made up of basalt, the same type of volcanic rock that makes up the Hawaiian Islands. This material originates within the deeper mantle region of Earth; it ascends along the rising zones of the convection cells. As this hot, dense mantle material rises toward the surface, it moves into regions of ever lessening pressure, because the weight of overlying material decreases. A lower-density liquid separates from the higher-density mantle material, rising to the surface as the "lava" we are familiar with from so many movies of erupting volcanoes. The magma enters a huge crack in the surface of the planet formed by the pulling apart of two plates and solidifies into basaltic ocean crust. It too begins to move away from the "spreading center" where it first lithified, and more new magma wells up to take its place—an endless conveyor belt.

The basalt produced in the spreading centers has a much different composition from its "parent," the mantle material rising along the limbs of the convection cells. Because it contains a much higher percentage of silica atoms, it is much lower in density than the mantle material. The basalt has *differentiated* from the parent material (which, when occasionally found on the surface, has the name peridiotite). This differentiation from a peridiotite composition to a basaltic composition is the final step of oceanic crust formation. Continents, however, have an even lower density than the oceanic crust. The recipe for their creation requires a further step in this arcane lithic cooking: the formation of the rock types granite and andesite. The characteristic speckled appearance of both of these rocks, compared to the more somber, chocolate to black color of basalt, comes from their containing even more of the white (and low-density) silica. The major step in forming continental crust is thus the differentiation of granite from material of a basaltic composition. This process takes place in several steps, but the key ingredient is water, and the key mechanism is called subduction.

Over many millions of years, oceanic crust moves away from its birthplace, the spreading centers, all the while being carried piggyback on the convecting mantle beneath it. Like all journeys, however, this long ride must eventually end; the oceanic crust cannot expand forever. The basalt has

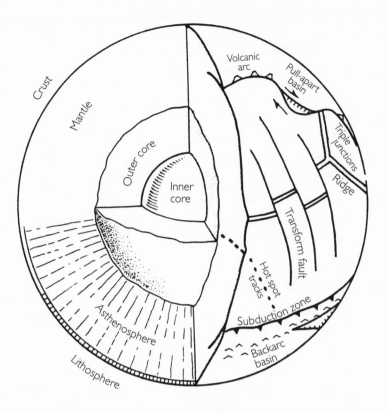

Figure 9.1 *Schematic diagram displaying the principal features of the lithosphere, the rigid plates 50–150 km thick that incessantly move about on Earth's surface, created along the 56,000 km of spreading-ridge systems and consumed along the 36,000 km of subduction zones.*

cooled through time, and even more significantly, it has gained some heavy freeloaders—piles of dense igneous rock known as gabbro that attach to the base of the basalt. The basalt now just barely floats, and as it cools, it gets heavier. Given any good excuse, it simply sinks, descending as deep as 650 kilometers. Eventually, then, the convection cell begins its downward journey back into the deep mantle, and when it does, it carries its veneer of oceanic crust back down with it, at regions called subduction zones.

Subduction zones (see Figure 9.1) are long, linear regions where oceanic crustal material is driven deep into Earth, not so much by being *pushed* down as sinking down through gravity. It is near and parallel to these subduction zones that linear mountain ranges are constructed. The mountains form partly as a by-product of the collision of two plates, which causes buckling and crumpling of the leading edges, and partly by the upward movement of hot magma, which eventually solidifies into granites and other magmatic rocks parallel to the subduction zones. The Cascade Mountains of Washington State are an example; the still-active peaks, such as Mt. Baker, Mt. Rainier, and Mt. St. Helens, are direct evidence of the power and importance of subduction in creating mountain ranges. Most of the world's volcanoes and mountain chains are found near these subduction zones (or where ancient subduction zones used to operate), further testimony to the fundamental link between subduction and mountain building. That mountain chains are not found on other planets or moons of our solar system is clear evidence that only Earth now has plate tectonics.

Volcanoes occur along subduction zones because by the time (which may be millions of years after its formation) that oceanic crust reaches a subduction zone and begins to descend, it is of slightly different composition from when it was created in the spreading centers. As the basalt created in spreading centers moves away from its birthplace, water is gradually added to the crystal structures of key minerals—in other words, the basalt becomes hydrated. Over long millennia, seawater works its way down through many cracks and crevices of the oceanic crust and reacts chemically through the addition of water molecules to the crustal lattices of minerals making up the basalt. Water-poor minerals actually incorporate significant amounts of water in their structure. The newly hydrated minerals have a lower melting point than nonhydrated minerals, so as the oceanic basalt descends in the subducting slab, the hydrated, silicate-rich minerals making up the basalt melt, and the liquid that is produced rises back toward the surface. This water leads to a decrease in the melting temperature of the overlying mantle rock that now surrounds it, creating liquid magma where one would otherwise expect to find only solid rock. This magma, when eventually cooled, becomes the rocks we call andesite and granite, and its rise back to the surface is a key force in producing new mountains and the line of volcanoes we find along subduction zones. But the crucial aspect of these volcanoes is that they are made up of magma of lower density than the basalt that parented them, and in this way, a new, lower-density rock type is created. This rock starts out as andesite (named after the Andes Mountains) and becomes part of the continental crust. Because andesite and granite (which is created in similar fashion) are so rich in silicate mineral, they are less dense than basalt. They become the backbone of the continents—and their flotation devices! With andesite- and granite-rich cores, continents can float on a sea of basalt. They can never be sunk in subduction zones. Continents cannot be destroyed (though they can be eroded). They can be split and fragmented, to drift from place to place, but their basic volume cannot be reduced. Through time, in fact, the number of continents on Earth has seemingly increased.

One of the most important findings about Earth history is that since the formation of our planet, the total area of oceanic plates has gradually diminished as the area of continental plates has grown (see Figure 9.2). This seems counterintuitive, because the oceans are continuously enlarging as a result of sea floor spreading. Yet as we have just seen, ocean crust *can* sink (and be remelted back to magma in the process), whereas the lighter continental crust remains afloat like a cork on this sea of basalt. Furthermore, the continents enlarge through the process of mountain building, for the volcanoes lining subduction zones and many continental edges receive vast quantities of granitic and andesitic magma. Geologist David Howell, in his book *Principles of Terrane Analysis*, estimates that the volume of continents increases by between 650 and 1300 cubic kilometers of rock each year. This estimate is for the modern day, and some geologists believe continental volume increased more rapidly in the past, especially early in Earth history, when plate tectonic processes may have occurred much faster than they do now because more heat emanated from the early Earth.

Plates thus intersect with each other in three ways: at the spreading centers (where new magma reaches the surface along enormous linear cracks, such as the mid-Atlantic ridge); areas where plates grind by each other side by side (such as the San Andreas Fault of California); and regions where

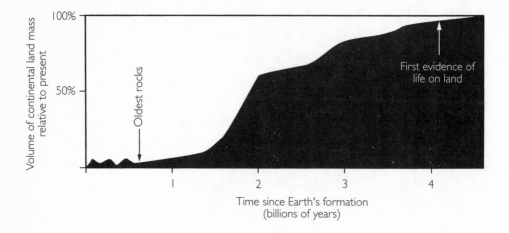

Figure 9.2   *Estimate of the growth of continental land mass with time (adapted from Taylor, 1999). For nearly a third of its history, Earth was a "water world" nearly devoid of land.*

plates collide—the subduction zones—which are associated with linear chains of active volcanoes (such as the Cascades and the Aleutian Islands).

WHY IS PLATE TECTONICS
IMPORTANT TO LIFE?

The rate of continental growth is of major importance to life and its ecosystems. The majority of Earth's biodiversity is today found on continents, and there is no reason to believe that this relationship has changed over the last 300 million years. As continents have grown through time, they have affected global climate, including the planet's overall albedo (its reflectivity to sunlight), the occurrence of glaciation events, oceanic circulation patterns, and the amount of nutrients reaching the sea. All of these factors have biological consequences and affect global biodiversity.

202

In the last chapter we proposed that diversity (roughly the number and relative abundance of species on the planet at any given time) is a major hedge, or defense, against planetary extinction or sterilization of life: High levels of diversity can counter the loss of body plans during mass extinctions. Plate tectonics can augment diversity by increasing the number and degree of separation of habitats (which promotes speciation). For example, as continents break apart, the seaways forming between them create barriers to dispersal. This in turn reduces gene flow and enhances the formation of new species through geographic isolation. Plate tectonics also increases the nutrients available to the biosphere, which may (or may not) also promote increased biotic diversity.

Plate tectonics promotes environmental complexity—and thus increased biotic diversity—on a global scale. A world with mountainous continents, oceans, and myriad islands such as those produced by plate tectonic forces is far more complex, and offers more evolutionary challenges, than would either totally land- or ocean-dominated planets without plate tectonics. James Valentine and Eldredge Moores first pointed out this relationship in a series of classic papers in the 1970s. They showed that changes in the position and configuration of the continents and oceans would have far-reaching effects on organisms, causing both increased diversification and extinction. Changes in continental position would affect ocean currents, temperature, seasonal rainfall patterns and fluctuations, the distribution of nutrients, and patterns of biological productivity. Such varying conditions would cause organisms to migrate out of the new environments—and would thus promote speciation. The deep sea would be least affected by such changes, but the deep sea is the area on Earth today with the fewest species. Over two-thirds of all animal species live on land, and the majority of marine species live in the shallow-water regions that would be most affected by plate tectonic movements.

The most diverse marine faunas on Earth today are found in the tropics, where communities are packed with vast numbers of highly specialized species. In higher latitudes the number of species lessens, and in Arctic regions there may be only a tenth as many species as in equivalent water depths or habitats found in the tropics. Not only are there *fewer* species in the higher latitudes, but the composition of species is also different there. Physiological adaptation constrains most species to fairly narrow temperature limits: Animal species adapted to warm, tropical conditions cannot survive in the cold, nor can the cold-adapted species tolerate the warmth of the tropics. Given that temperature conditions change rapidly with latitude, it's not surprising that north-south coastlines of continents show a continuously changing mix of species. North-south coastlines promote diversity because of the latitudinal temperature gradients. East-west coastlines, on the other hand, often show similar species.

As continental positions change through time, the relative abundance of north-south and east-west coastlines can change. Also, the larger the continents, the lower the environmental heterogeneity. If many or all of the continents were welded together into "supercontinents," biodiversity would be expected to be lower than if there were many smaller and separated continental masses. On one large continent, groups of land animals have fewer barriers to dispersal—and thus less opportunity to form new species. Clearly, continental size and position should affect biodiversity, and this appears to have been the case in Earth history.

WHAT WOULD HAPPEN IF PLATE TECTONICS CEASED?

The fossil record suggests that there are more species of animals and plants alive on Earth today than at any time in the past; estimates vary between about 3 and 30 million species. This great diversity has come about through many physical and evolutionary factors. We contend that the effects of plate tectonics are among the most important. But once created, does high biodiversity require the *continued* presence of plate tectonics? We can examine this question with a thought experiment.

An End to Volcanism

Imagine that all volcanism on Earth's surface suddenly ceases. This will stop the many dozens of volcanic eruptions that occur on the continents each year (usually causing great media fanfare and little damage). But the cessation of volcanism will have a far more profound effect. If all volcanism stops, so does sea floor spreading—and thus plate tectonics as well. And if plate tectonics stops, Earth eventually (through erosion) loses most or all of the continents where most terrestrial life exists. In addition, CO_2 is removed from the atmosphere via weathering, causing our planet to freeze. Of all of the attributes that make Earth rare, plate tectonics may be one of the most profound and—in terms of the evolution and maintenance of animal life—one of the most important.

Only cessation of the flow of heat up from Earth's interior or thickening of the crust would stop volcanism. It is this heat that causes convective motion of the interior, the subterranean motor of plate tectonics. To stop plate tectonics would require eliminating these great lithic boiling pots, and that cannot be done unless all heat emanating from the Earth's interior is stopped (which would require that all the radioactive minerals locked away there decay to stable daughter products) or the composition of Earth's crust or upper mantle changes such that movement can no longer occur. This could happen if the crust became too thick or the mantle too viscous to allow movement. None of these conditions is likely to occur on Earth in the foreseeable future, but there is speculation that just such events occurred on both Venus and Mars in the past.

If Earth's tectonic plates did suddenly stop moving, subduction would no longer occur at the contacts between colliding plates. Mountains—and mountain chains—would cease to rise. Erosion would begin to eat away at their height. Eventually, the world's mountains would be reduced to sea level. How long would it take? The problem is a bit more complicated than simply measuring average erosion rates and calculating the number of years required for the mountains to disappear. This is because of the principle of isostacy. Mountains (and continents) are a bit like icebergs: If you cut off the top, the bottom rises up relative to sea level, causing the entire iceberg (or mountain) to rise. Eventually, however, even this isostatic rebound effect would be overcome by the extent of the erosion.

What would sea level be in a world without further plate tectonics? All of the sediment produced by the simultaneous erosion of the world's mountains would have to go somewhere—and that somewhere is the ocean. The eroding continental mass carried into the oceans by river and wind transport would displace seawater and cause the level of the sea to rise. Calculations by David Montgomery, a geomorphologist at the University of Washington, suggest that the entire Earth might become covered by a global ocean—much shallower than the oceans of today, of course, but global in extent nevertheless. Our planet would have returned to its state of 4 billion years ago: a globe covered completely (or nearly so) by ocean. And with the continents awash, Earth would witness a mass extinction more catastrophic than any in the past. All land life would die off under the lapping waves. Paradoxically, the increase of ocean area would probably also be accompanied by extinctions in the sea. Ocean life depends on nutrients, and most nutrients come from the land as runoff from rivers and streams. With the disappearance of land, the total amount of nutrients (though initially higher as so much new sediment entered the ocean system) would eventually lessen, and with fewer resources, there would be fewer marine animals and plants.

How long before such a water world would be achieved? Tens of millions of years would be required for the mountains and continents to erode to sea level. Yet mass extinction would ensue long before that. Planetary calamity for complex life would occur shortly after the cessation of plate movement, for plate tectonics is not only the reason we have mountains; it turns out to control our planet's climatic thermostat as well.

Loss of Planetary Temperature Control

The temperature of Earth must remain in a range suitable for the existence of liquid water if animal life is to be maintained. The range of temperature that Earth experiences is the result of many factors. One is the existence of the atmosphere. The average temperature of the Moon is $-18°C$, for example, well below the freezing point of water, simply because it has no appreciable atmosphere. If Earth did not have its cloaking atmosphere, including such insulating gases as water vapor and carbon dioxide (producing the much-discussed Greenhouse Effect), its temperature would be about the same as that of the Moon. Yet the Earth, thanks to the greenhouse gases, has an average global temperature of 15°C (33°C warmer than the Moon). Greenhouse gases are keys to the presence of fresh water on this planet and thus are keys to the presence of animal life—and many scientist now believe that the balance of greenhouse gases in Earth's atmosphere is directly related to existence of plate tectonics.

Greenhouse gases are those with three or more atoms, such as water vapor (H_2O; three atoms), ozone (O_3), carbon dioxide (CO_2; three atoms), and methane (CH_4; five atoms). All can capture outgoing infrared energy from Earth's surface and, in so doing, warm the planet. Their role in keeping Earth's temperature within the critical levels necessary not only for allowing the presence of liquid water (0°C to 100°C) but also for maintaining animals (about 2°C to about 45°C) has been beautifully summarized by Columbia University geologist Wally Broecker in *How to Build a Habitable Planet.* Broecker describes the following scenario. Imagine that the sun's energy was diminished for a period of time brief by geological standards but long enough for the oceans to freeze. If the sun then resumed its normal output of today, Earth would remain frozen. Once frozen, water reflects much of the light that hits it, and even the current volume of greenhouse gas would be insufficient to reheat the planet to a temperature at which the water would thaw. This condition is called a Global Icehouse, and it is one way a planet can lose its animal life. They freeze to death.

Now, say we reversed this situation and allowed the sun's energy to *increase* for a geologically short period of time, but long enough so that all of Earth's oceans boiled away, filling the atmosphere with steam. If we then reduced the sun's energy to its present-day levels, the oceans might not recondense, and the planet would stay hot. Once in the atmosphere, the steam would keep the planet hot through its properties as a greenhouse gas, even when solar radiation hitting the planet had decreased. This situation is called a Runaway Greenhouse.

The Earth's greenhouse gases are rare compounds of our planet's atmosphere. It turns out that the major constituents of our atmosphere, nitrogen and oxygen, play little role in the greenhouse warming, because they do not absorb infrared radiation. Carbon dioxide and water vapor, on the other hand, do, even though they make up only a tiny fraction of the gas volume of the atmosphere (carbon dioxide constitutes only 0.035% of the atmosphere). Plate tectonics plays an important part—perhaps the most important part—in maintaining levels of greenhouse gases, and these in turn maintain the temperatures necessary for animal life.

PLATE TECTONICS AS GLOBAL THERMOSTAT

Over and over again we come back to to a common theme: the importance of liquid water. For animal life based on DNA to exist and evolve, water must be present and abundant on a planet's surface. Even on the water-rich Earth today, slight differences in water content obviously affect life. In desert regions there is little life; in rainforests at the same latitude, life teems in abundance. For complex life to be attained (and then maintained), a planet's water supply (1) must be large enough to sustain a sizable ocean on the planet's surface, (2) must have migrated to the surface from the planet's interior, (3) must not be lost to space, and (4) must exist largely in liquid form. Plate tectonics plays a role in all four of these criteria.

Earth is about one-half of 1% water by weight. Much of this water arrived among the planetesimals that took part in Earth's formation and accretion. Other volumes of it were dumped here by incoming comets after Earth accreted. The relative importance of these two processes is largely unknown at this time.

Once liquid water is established on the surface of a planet, its maintenance becomes the primary requirement for attaining (and then support-

208

ing) animal life. The maintenance of liquid water is controlled largely by global temperatures, which are a by-product of the greenhouse gas volumes of a planet's atmosphere. The temperature of Earth's (and of any planet's surface) is a function of several factors. The first is related to the energy coming from its sun. The second is a function of how much of that energy is absorbed by the planet (some might be reflected into space, and this relationship is dictated by a planet's reflectivity, or albedo). The third is related to the volume of "greenhouse gases" maintained in a planet's atmosphere. Greenhouse gases have a residence time in any atmosphere and are eventually broken down or undergo a change in phase. If their supplies are not constantly replenished, the planet in question (such as Earth) will grow colder gradually until the freezing temperature of water is reached, at which point it will grow colder *rapidly* (as we have noted, when a planet starts accumulating ice, its albedo increases, boosting its rate of cooling). Greenhouse gases are thus vastly important in maintaining a planet's thermostatic reading. Both plate tectonic and non–plate-tectonic planets regularly produce greenhouse gases, because the most important source of these planetary insulators is volcanic eruption, which occurs on most or all planets. On Earth, volcanoes daily exhale vast volumes of carbon dioxide from deep within. Even so-called "dormant" volcanoes are venting carbon dioxide into the atmosphere. On any planet with volcanism there is usually an abundance of greenhouse gases—too much in some cases, and this is where plate tectonics becomes crucial.

Greenhouse gas compositions, and thus planetary temperatures, are by-products of complex interactions among a planet's interior, surface, and atmospheric chemistry. One of the most important by-products of plate tectonics is the recycling of mineral and chemical compounds locked up in any planet's sedimentary rock cover. On non–plate-tectonic worlds, vast quantities of sedimentary material are produced by erosion. These materials and minerals become sequestered and eventually buried and lithified through sedimentation and the formation of sedimentary rocks, and in most cases, they are re-exhumed only through some process leading to mountain building. Yet, as we have seen, mountain building on non–plate-tectonic worlds is largely confined to the formation of large volcanoes over hot spots. With plate tectonics, however, the motion (and collision) of plates, the formation of mountain chains, and the process of subduction all lead to a recycling of many materials. This recycling plays a large role in maintaining Earth's global temperature values in a range that allows the existence of liquid water. One of the most important of the recycling processes is putting CO_2 back into the atmosphere. As limestone is subducted deep into the mantle, it metamorphoses and, in the process, returns CO_2 into the atmosphere. This is clearly an important aspect of global warming.

The most important element in *reducing* atmospheric carbon dioxide (which leads to global cooling) is the weathering of minerals known as silicates, such as feldspar and mica (granite has many such minerals within it). The presence or absence of plate tectonics on a given planet greatly affects the rates and efficiency of this "global thermostat." The basic chemical reaction is $CaSiO_3 + CO_2 = CaCO_3 + SiO_2$. When the first two chemicals in this equation combine, limestone is produced and carbon dioxide removed from the system. The feedback mechanism at work here was first pointed out in a landmark 1981 paper by J. Walker, P. Hays, and J. Kasting. (James Kasting has told us that he first had this insight in the middle of his Ph.D. exam!) The mechanism is related to the rates of weathering—that is, the physical or chemical breakdown of rocks and minerals. Although weathered entails the reduction in size of rocks (big boulders weather into sand and clay over time), a very important chemical aspect is also involved (see Figure 9.3). Weathering can cause the actual mineral constituents of the rocks being weathered to change. Weathering of rocks that contain silicate minerals (such as granite) plays a crucial part in regulating the planetary thermostat. Walker and his colleagues pointed out that as a planet warms, the rate of chemical weathering on its surface increases. As the rate of weathering increases, more silicate material is made available for reaction with the atmosphere, and more carbon dioxide is removed, thus causing *cooling*. Yet as the planet cools, the rate of weathering decreases, and the CO_2 content of the atmosphere begins to rise, causing *warming* to occur. In this fashion the Earth's temperature oscillates between warmer and cooler as a result of the carbonate–silicate weathering and

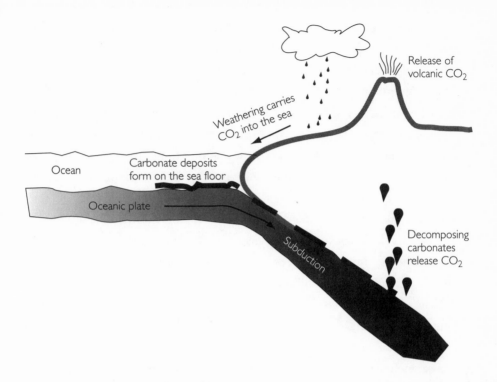

Figure 9.3   *The CO_2–rock weathering cycle. This remarkable cycle has controlled the amount of atmospheric carbon dioxide, a greenhouse gas, to regulate Earth's surface temperature for billions of years. Because this process requires both surface water and plate tectonics, it is not known to occur elsewhere.*

precipitation cycles. Without plate tectonics, this system does not work efficiently. It also works less efficiently on planets without land surfaces—and *much* less efficiently on planets without vascular plants such as the higher plants common on Earth today.

Calcium is an important ingredient in this process, and it has two main sources on a planet's surface. It is found in igneous rocks and (more important) in the sedimentary rocks called limestone. Calcium reacts with carbon dioxide to form limestone, the material that marine animals use to build their shells (and that we humans use to build our cement and concrete). Calcium thus draws CO_2 out of the atmosphere. When CO_2 begins to increase in the atmosphere, more limestone formation occurs, but only if there is a steady source of new calcium available. The calcium content is steadily made available by plate tectonics, for the formation of new mountains brings new sources of calcium back into the system by exhuming (in magmas) ancient limestone, eroding it, and thus releasing its calcium to react with more CO_2.

The planetary thermostat requires a balance between the amount of CO_2 being pumped into the atmosphere through volcanic action and the amount being taken out through the formation of limestone. On non–plate-tectonic worlds, buried limestone stays buried, thus removing calcium from the system and producing increases in carbon dioxide. On Earth, at least, plate tectonics plays an integral part in maintaining a stable global temperature by recycling limestone into the system.

Although most accounts of habitability of planets refer to the range between 0°C and 100°C, required temperature range is really much narrower if animals are to survive. As we have seen, life such as bacteria can withstand a range of temperatures that may approach 200°C in high-pressure environments. But animals are much more fragile. Animal life on Earth—and perhaps anywhere in the Universe—depends on the narrowest of temperature ranges within the wider range that permits liquid water to exist. Extended periods of anything above 40°C or much below 5°C will stymie animal life. The planetary thermostat must be set to a narrow range of temperatures indeed, and it may be that only the plate tectonic thermostat makes this fine-tuning possible.

PLATE TECTONICS AND THE MAGNETIC FIELD

Outer space is not a particularly friendly place. One of its hazards is cosmic rays, which are elementary particles—electrons, protons, helium nuclei, and heavier nuclei—traveling at velocities approaching the speed of light. They come from many sources, including the sun and cosmic rays from distant supernovae, the explosions of stars. These catastrophic events send great numbers of particles hurtling through space.

In *The Search for Life in the Universe*, D. Goldsmith and T. Owen speculate that without some sort of protection, life on Earth's surface would be extinguished within several generations by cosmic rays hitting our planet's surface. However, the vast majority of cosmic rays are deflected by Earth's magnetic field. The innermost layer of our planet, its core, is made up mainly of iron, which in the outermost region of the core is in a liquid state. As Earth spins, it creates convective movement in this liquid that produces a giant magnetic field surrounding the entire planet. What produces the convection cells in the core is loss of heat. Heat must be exported out of the core, and this liberation of heat appears to be greatly influenced by Earth's plate tectonic regime. Joseph Kirschvink of Cal Tech has suggested that without plate tectonics, there would not be enough temperature difference across the core region to produce the convective cells necessary to generate Earth's magnetic field; no plate tectonics, no magnetic field. The magnetic field also reduces "sputtering" of the atmosphere, a process whereby the atmosphere is gradually lost into space. No magnetic field, perhaps no animal life. Plate tectonics to the rescue again.

WHY DOES EARTH (BUT NOT MARS OR VENUS) HAVE PLATE TECTONICS?

Why is there plate tectonics on Earth? The recipe for plate tectonics seems simple enough at first glance. You need a planet differentiated into a thin, solid crust sitting atop an underlying region that is hot, fluid, and mobile. You need this underlying region to be undergoing convection, and for that you need heat emanating from even deeper in the planet. And you are likely to need water—oceans of water: Much new research suggests that without water you cannot have plate tectonics (though perhaps it is simply that without water you cannot get continents).

As in so much else in planetary geology, there is still a great deal we don't know about why our planet (and, more important, any planet) develops and then maintains plate tectonics. Because ours is still the only planet we know that has plate tectonics, we have nothing with which to compare it.

Much of the data pertaining to plate tectonics lies so deep that we are unlikely ever to sample it directly.

As an illustration of the degree of uncertainty about what we might call planetary plate tectonics, which we can define as the theoretical (as opposed to the Earth-based actual) study of plate tectonics, we cannot be certain whether plate tectonics would operate if Earth were 20% larger or smaller, or if it had a crust with more iron and nickel than it does, or if its surface had only 10% of the present-day volume of water. The best current work on these types of questions is being done by planetary geologists V. Solomatov and L. Moresi, who are using computational models to study how convection (the driving force of plate tectonics) works. Yet the abstract of their 1997 paper on the subject concluded, "The nature of the mobility of lithosphere plates on Earth has yet to be explained." We know the plates move, and we know convection moves them. The physics behind the convection is well understood, but its application to subduction is still an enigma.

When we asked about the physical condition necessary to produce plate tectonics on a planet, Solomatov responded, "It is a very interesting problem and we've just started exploring the physical conditions required for plate tectonics to occur on a planet. So far, we have been moving to the conclusion that water might be the factor which is crucial for plate tectonics: no water, no plate tectonics." Without water, the lithosphere (which is the *plate* of plate tectonics, the rigid surface region composed of the crust and uppermost part of the mantle) is strong and cannot break and descend back into the mantle— the process known as subduction that occurs along the linear subduction zones described earlier in this chapter. According to Solomatov, subduction is a major requirement for plate tectonics. Apparently, subduction zones operate only when the crust is "weak," or able to bend and break, which allows it to descend into the regions where the mantle convection cells sink downward. All of this work is being done with mathematical modeling. Solomatov and his colleagues are using computers to arrive at these generalizations—not trips to the center of the Earth with Jules Verne's heroes.

Even in the absence of water, plumes of hot magma may rise to a planet's surface. But this new material must ultimately go somewhere, and if subduction is not operating, the plates will not move, for the new crustal material must ultimately duck down into the mantle, along the linear subduction zones. Without subduction zones, there is no plate tectonics, even if mantle convection cells are operating inside a planet.

Venus and Mars both lack subduction zones and thus lack plate tectonics. Although both might have the internal mantle convection necessary to move surface plates, the surface itself is composed of "strong" rock (Solomatov's term) that cannot move. Because of its thickness and strength, the crust on these planets is now immobile. The lack of water on both of these plates may be the reason why this is so. Because both of these planets may in the past have had liquid water and crustal composition similar enough to that of Earth, we may find that Venus and Mars once did have plate tectonics—and perhaps lost it when they lost their liquid water. Venus and Mars may be experiencing what Solomatov and Moresi describe as a "stagnant lid regime": The viscosity difference between the convecting mantle and the solid surface is so great that little or no movement of the crust can occur. Yet heat continues to flow upward, and in the case of Venus, this heat caused the entire surface of the planet to melt about a billion years ago (the planetary "resurfacing" we alluded to at the start of this chapter). On Earth this great viscosity difference does not occur. Earth has a "small viscosity contrast regime," according to the technical scientific papers describing all of this, and the result is the very actively moving crust so important for mountain formation, nutrient cycling, and life.

Yet perhaps we have this story reversed. Perhaps Mars and Venus had water but lost it because they had no plate tectonics—and thus no planetary thermostat.

HOW (AND WHEN) DID PLATE TECTONICS START ON EARTH?

The time of onset of plate tectonics is controversial. Many sources believe it began 1 to 2 billion years after Earth's formation, whereas others view plate tectonics as being far more ancient, its inception dating back over 4 billion years. Much of the controversy involves the rate of heat flow from the early Earth and how this would have affected the composition and rigidity of the planet's surface.

By the time the crust had solidified, more than half of the heat that could result from planetary accretion, core formation, and decay of radioactive isotopes (such as uranium-235) had already been lost from Earth. During the 2 billion years of the Archaean era, heat flow slowed. Some workers believe that the early crust was still too hot and thin to act as the rigid plate necessary for plate tectonics; according to this hypothesis, plate tectonics may not have commenced until 2.5 billion years ago. There is evidence in much older rocks, however, of fault lines and movements consistent with plate tectonics.

The rate at which plate tectonics built continental surfaces on Earth was not constant. If we plot the size of the continents through time relative to the present area, we do see not a linear increase but a logistic curve—a curve that started slowly, picked up speed in its middle, and then slowed near the end. We have spoken in another context of the "Cambrian Explosion." Here, Earth underwent a "continental explosion" that resulted in a rapid formation of land area. Many lines of evidence suggest that by far the greatest growth took place rather rapidly, during a period between about 2 and 3 billion years ago. This rapid growth completely changed Earth from a planet dominated by oceans to one dominated (at least in terms of its global temperatures and chemistry) by continents.

Could Plate Tectonics Actually Have Inhibited the Formation of Animal Life on Earth?

In this chapter we have contended that plate tectonics facilitated the rise and then the maintenance of animal life on Earth. But might not the opposite actually be true? Could it be that plate tectonics actually retarded the rise of animals? This is the contention of two NASA scientists, H. Hartman and C. McKay, who hypothesized that plate tectonics slowed the rate of oxygenation of the Earth atmosphere. In a 1995 article, Hartman and McKay proposed that plate tectonics slowed the rise of oxygenation on Earth and, by inference, on any planet.

As we have detailed in an earlier chapter, animal life did not arise on Earth until less than a billion years ago, whereas life on this planet antedates the first animals by about 3 billion years. One of the most puzzling aspects of life's history on Earth is this singular gap between the first life and the first animal life. Many factors were surely involved, but there is irrefutable evidence that oxygen is a necessary ingredient for animal life (at least on Earth), and there is much evidence that sufficient concentrations of oxygen were not present in the oceans and atmosphere until less than 2 billion years ago. Many scientists suspect that the long time it took for Earth to acquire an oxygen atmosphere accounts for some, or even all, of the delay between the origin of the first life and the origin of animal life on Earth. Hartman and McKay make the novel suggestion that this delay was partly due to the existence of plate tectonics on Earth.

It is universally agreed that the rise of oxygen on Earth was due to the release of free oxygen as a by-product of photosynthesis. The earliest photosynthetic organisms used an enzymatic pathway called Photosystem 1; however, this system does not release free oxygen. The later-evolved Photosystem 2 does. This latter system may not have evolved until 2.7 to 2.5 billion years ago. Eventually, photosynthesizing organisms such as photosynthetic bacteria and single-celled plants floating in the early seas would have released vast volumes of oxygen. There was probably some source of inorganically produced free oxygen on the early Earth as well. It may be, for example, that ultraviolet rays hitting water vapor in the upper atmosphere created free oxygen, at least in small volumes. However, a net accumulation could not take place until various reducing compounds (which bind the newly released oxygen and keep it from accumulating as a dissolved gas in the oceans or as a gas in the atmosphere) were used up. For example, the amount of iron in the crust

217

of a planet has a major effect, for all of it on the surface in contact with the atmosphere must be oxidized before free oxygen can accumulate. Such reducing compounds emanate from volcanoes, and it can be argued that planets with a higher rate of volcanicity have more reducing compounds in their oceans and atmospheres. Another important source of reducing compounds is organic compounds, produced either through the death and rotting of organisms or through the inorganic formation of organic compounds, such as amino acids. Great volumes of such material are found in the oceans on Earth, but it is usually buried in sediments. In the absence of plate tectonics, argue Hartman and McKay, such sediments become buried in sedimentary basins and are never brought back into contact with the oceans and atmosphere; thus they are removed from active participation in oceanic and atmospheric chemistry. Because they are taken out of the system, oxygen can accumulate faster than in the case where reducing compounds are constantly being reintroduced into the atmosphere—a case where the dead don't stay buried.

Hartman and McKay make the intriguing point that Mars may have seen the evolution of complex life within 100 million years of the formation of that planet (assuming, of course, that life originated there at all). Their argument is as follows: The rapid removal of reductants on Mars through burial in deep and undisturbed sediment would have allowed oxygenation to occur much more quickly than on Earth (see Figure 9.4), where plate tectonics constantly recycles sediments via subduction, plate collision, and mountain building. All of these processes can cause previously buried sediments to be brought back up to the surface, where their reductants would once more bind whatever atmospheric oxygen was available. Hartman and McKay also point out that volcanicity on a planet like Mars that does not exhibit plate tectonics is much lower than on Earth. Thus the amount of reducing compounds (such as hydrogen sulfide) entering the atmosphere–ocean systems on Mars from volcanic sources would also have been much lower.

Could it be, then, that Earth hosted the evolution and then the maintenance of animal life *in spite of* plate tectonics? And that plate tectonics actually discourages the attainment of animal life on a given planet because its presence slows the accumulation of the necessary oxygen-rich atmosphere?

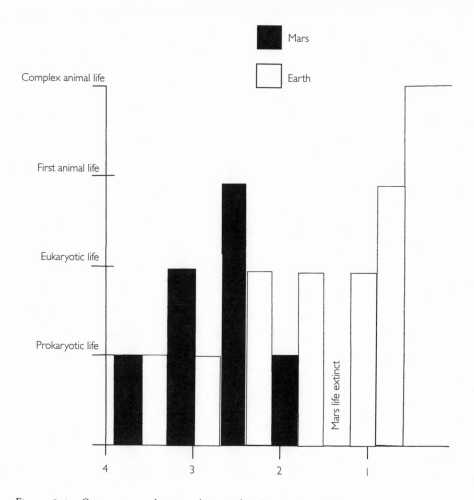

Figure 9.4 *Comparative evolutionary history of a planet without plate tectonics (Mars) and a world with plate tectonics (Earth) expressed in billions of years before the present.*

We cannot fault the arguments of Hartman and McKay concerning the role of reductants in retarding oxygenation. However, we can point out that plate tectonics would surely increase the rate of biologically produced oxygen on any world, because it enhances biological productivity by recycling nutrients such as nitrates and phosphates. The net productivity on plate tectonic worlds should thus be expected to be far higher than on non–plate-tectonic worlds, so the rate of oxygenation through photosynthesis should also be much higher on a plate tectonic world, perhaps offsetting the retardant effect of the recycling of sediment-sequestered reductants.

MOST CRUCIAL ELEMENT OF THE RARE EARTH HYPOTHESIS?

Plate tectonics plays at least three crucial roles in maintaining animal life: It promotes biological productivity; it promotes diversity (the hedge against mass extinction); and it helps maintain equable temperatures, a necessary requirement for animal life. It may be that plate tectonics is the central requirement for life on a planet and that it is necessary for keeping a world supplied with water. How rare is plate tectonics? We know that of all the planets and moons in our solar system, plate tectonics is found only on Earth. But might it not be even rarer than that? One possibility is that Earth has plate tectonics because of another uncommon attribute of our planet: the presence of a large companion moon, the subject of the next chapter.

The Moon, Jupiter, and Life on Earth

The surface is fine and powdery. I can kick it up loosely
with my toe. It does adhere in fine layers, like powdered
charcoal, to the sole and sides of my boots. I only go in a
small fraction of an inch, maybe an eighth of an inch, but I
can see the footprints of my boots and the treads in the
fine, sandy particles.

—Neil Armstrong's first words from the surface
of the Moon (1969)

Perhaps an astronomer's greatest fear is that sooner or later, someone will
mistake him or her for an astrologer. The ancient belief system known
as astrology posits that the stars and planets exert a major influence on
our daily lives—a set of beliefs frequently and fervently disputed by astronomers. Recent research, however, has in an odd way proved the as-
trologers slightly correct. Two heavenly bodies, the Moon and Jupiter, do, in
fact, play pivotal roles in our very existence as a species. Without the Moon,
and without Jupiter, there is a strong likelihood that animal life would not

exist on Earth today. Both are thus key elements in the Rare Earth Hypothesis, but for different reasons.

LUNA

Without the Moon there would be no moonbeams, no month, no lunacy, no Apollo program, less poetry, and a world where every night was dark and gloomy. Without the Moon it is also likely that no birds, redwoods, whales, trilobites, or other advanced life would ever have graced Earth.

Although there are dozens of moons in the solar system, the familiar ghostly white moon that illuminates our night sky is highly unusual, and its presence appears to have played a surprisingly important role in the evolution of life. The Moon is just a spherical rock 2000 miles in diameter and 250,000 miles away, but its presence has enabled Earth to become a long-term habitat for life. The Moon is a fascinating factor in the Rare Earth concept because the likelihood that an Earth-like planet should have such a large moon is small. The conditions suitable for moon formation were common for the outer planets but rare for the inner ones. Of the many moons in the solar system, nearly all orbit the giant planets of the outer solar system. The warm, Earth-like planets that are close to the sun and that fall within the habitable zone, are nearly devoid of moons. The only moons of the terrestrial planets are ours and Phobos and Diemos, the two tiny (10 kilometers in diameter) moons of Mars. Some of the solar system's moons are huge. Jupiter's Ganymede is nearly as large as Mars, and Saturn's Titan is nearly that large and has an atmosphere denser than our own, though much colder. Our Moon is somewhat of a freak because of its large size in comparison to its parent planet. The Moon is nearly a third the size of Earth, and in some ways it is more of a twin than a subordinate. The only other case in the solar system where a moon is comparable in size to its planet is Pluto and its moon, Charon.

Tilt!

The Moon plays three pivotal roles that affect the evolution and survival of life on Earth. It causes lunar tides, it stabilizes the tilt of Earth's spin axis, and it slows the Earth's rate of rotation. Of these, the most important is its effect on the angle of tilt of Earth's spin axis relative to the plane of its orbit, which is called "obliquity." Obliquity is the cause of seasonal changes. For most of Earth's recent history, its obliquity has not varied by more than a degree or two from its present value of 23 degrees. Although the direction of the tilt varies over periods of tens of thousands of years as the planet wobbles, much like the precession of a spinning top, the angle of the tilt relative to the orbit plane remains almost fixed. This angle is nearly constant for hundreds of millions of years because of gravitational effects of the Moon. Without the Moon, the tilt angle would wander in response to the gravitational pulls of the sun and Jupiter. The monthly motion of our large Moon damps any tendencies for the tilt axis to change. If the Moon were smaller or more distant, or if Jupiter were larger or closer, or if Earth were closer to or farther from the sun, the Moon's stabilizing influence would be less effective. Without a large moon, Earth's spin axis might vary by as much as 90 degrees. Mars, a planet with the same spin rate and axis tilt, but no large moon, is believed to have exhibited changes to its tilt axis of 45 degrees or more.

Because tilt of a planet's spin axis determines the relative amounts of sunlight that land on the polar and on the equatorial regions during the seasons, it strongly affects a planet's climate. On planets with moderate tilts, the majority of solar energy is absorbed in the equatorial regions, where the noon sun is always high in the sky. Each pole is in total darkness for half a year and has constant illumination for half a year. The highest altitude that the sun reaches in the sky at the pole is exactly equal to the number of degrees of the tilt of the spin axis. For moderate tilt angles, the sun is never high in the polar sky, and ground heating by sunlight is low even in the middle of the summer. The planet Mercury provides a spectacular example of what can happen on a planet whose spin axis is nearly perfectly perpendicular to the plane of its orbit. Mercury is the closest planet to the sun and most of its surface is hell-ishly hot, but radar imaging from Earth has shown that the poles of the planet are covered with ice. The planet is very close to the sun, but as viewed from the poles, the sun is always on the horizon. In contrast to Mercury's lack of tilt, the planet Uranus has a 90-degree tilt; and one pole is exposed to sun-light for half a year while the other experiences cryogenic darkness.

Although our viewpoint is certainly biased, our planet's tilt axis seems to be "just right." Constancy of the tilt angle is a factor that provides long-term stability of Earth's surface temperature. If the polar tilt axis had undergone wide deviations from its present value, Earth's climate would have been much less hospitable for the evolution of higher life forms. One of the worst possi-bilities is that excessive axis tilt could have led to the total freezing over of the oceans, a situation that might be very difficult to recover from. Extensive ice cover increases the reflectivity of the planet, and with less absorption of sunlight, the planet continues to cool. Astronomer Jacques Laskar, who made many of the calculations that led to the surprising discovery of the Moon's importance in maintaining Earth's stable obliquity, summarized the situation as follows:

> These results show that the situation of the Earth is very peculiar. The common status for all the terrestrial planets is to have experi-enced very large scale chaotic behavior for their obliquity, which, in the case of the Earth and in the absence of the Moon, may have prevented the appearance of evolved forms of life. . . . [W]e owe our present climate stability to an exceptional event: the presence of the Moon.

High obliquity has remarkable and seemingly counterintuitive effects on planets (see Figure 10.1). Consider a planet that is tipped 90 degrees. Av-eraged over the year, the poles would receive exactly as much solar energy as the equator would with no tilt angle. The north pole would become the Sa-hara! For the 90-degree tilt, however, the equatorial regions would receive much less energy averaged over the year and would become colder. If a planet is tilted more than 54 degrees, its polar regions actually receive more

224

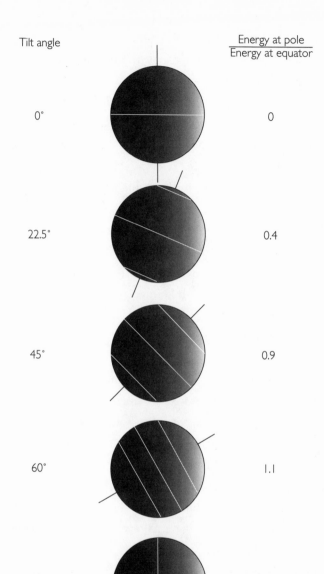

Tilt angle

$$\frac{\text{Energy at pole}}{\text{Energy at equator}}$$

Tilt angle	Energy at pole / Energy at equator
0°	0
22.5°	0.4
45°	0.9
60°	1.1
90°	1.6

Figure 10.1   *The ratio of the annual amount of solar energy falling on a planet's pole to that falling on its equator varies with the angle of a planet's spin axis. With a tilt angle of 22.5°, the Earth has very cold polar regions, but if the tilt exceeded 50°, the polar regions would actually receive more sunlight than the tropics. The lines parallel to the equator are the polar circles, where the Sun never sets in the midsummer and never rises in the midwinter.*

energy input from sunlight than the equatorial regions. If the Earth were tilted more than this amount, the equatorial oceans might freeze and the polar regions would be warmer: a topsy-turvy world. Recently uncovered evidence has revealed that equatorial ice sheets did exist about 800 to 600 million years ago, and ice-rafted sediments of this age have been found in formerly equatorial regions. This has led to the "Snowball Earth" hypothesis that Earth may have actually have frozen over, as we saw in Chapter 6. It has been suggested that this may have been due to high tilt angle during a period of time when the Moon did not have full control. We do not know for sure how long the Moon has been successful in stabilizing Earth's obliquity.

In the distant future, the Moon will lose its ability to stabilize Earth's spin axis. The Moon is slowly moving outward from Earth (at a rate of about 4 centimeters a year), and within 2 billion years it will be too far away to have enough influence to stabilize Earth's obliquity. Earth's tilt angle will begin to change as a result, and the planet's climate will follow suit. Further complicating the future is the slow but unrelenting increase in the brightness of the sun. At the time when our planet's spin axis begins to wander, the sun will be hotter, and both effects will decrease the habitability of Earth.

There is currently much speculation about how rapid such changes of planetary obliquity might be in the absence of the Moon. Estimates for the time it would take Earth to "roll" on its side range from tens of millions of years to far shorter periods. Astronomer Tom Quinn of the University of Washington has suggested to us that the time of obliquity change could occur on scales as short as hundreds of thousands, rather than millions, of years. Such large-scale fluctuations would probably lead to very rapid and violent climate change. If the tropical regions became locked in a permanent ice cover in 100,000 years or less, there would certainly be a mass extinction of great severity.

Is the lack of a large moon sufficient to prevent microbial life from evolving into animal life? We have no information, but because deep-sea regions are insulated from climate change, it seems doubtful that rapid obliquity changes would deprive a planet of animal life. What it could do, however, is deprive a planet of complex life on land.

Lunar Tides

A second benefit of Earth's large moon is tides, which are due to gravitational effects of both the sun and the Moon. The pull of these two bodies produces bulges in the ocean pointing both toward and away from the Moon and the sun. The complexity of Earth's present tidal effects is well illustrated by the tidal charts cherished by clam diggers, anglers, and sailors. The daily variations seen in the charts are caused by the interplay of both lunar and solar tides. Both the Moon and the sun cause ocean bulges on their respective near and far sides of the planet. As Earth spins under the bulges, the sea rises and falls at any particular location. When the Moon lines up with the sun every 2 weeks, the tidal ranges are at a maximum, and when they are 90 degrees apart in the sky (the quarter moon is overhead at sunrise or sunset), the range of tidal change is minimized. With a smaller or more distant moon, the lunar tides would be lesser and would have a different annual variation.

Soon after the Moon formed, it was perhaps 15,000 miles from Earth. Instead of being a few meters high, as they are today, it is possible that lunar tides rose hundreds of meters or higher. The extreme effects of such a close moon could have strongly heated Earth's surface. The ocean tides (and land tides) from a nearby Moon would have been enormous, and the flexing of Earth's crust, along with frictional heating, may have actually melted the rocky surface. However severe their effects, the enormous tidal variations would have been short-lived because the forces responsible for tides also cause the Moon to move outward, thus diminishing the effect. Early land tides may have been a kilometer high, but they dropped to moderate levels in less than a million years.

The retreat of the Moon is a natural consequence of gravitational pull between the Moon and the tidal bulges. The Earth's lunar tidal bulges don't actually correspond with a line from Earth to the Moon but, rather, lead ahead as the Moon orbits the planet (see Figure 10.2). This offset produces a torque that causes Earth's spin rate to decrease slowly and the distance be-

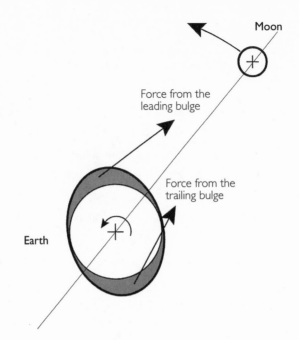

Figure 10.2 *Earth's leading tidal bulge produces a constant forward force on the Moon that is not totally balanced by the backward pull of the more distant trailing bulge. The net forward force on the Moon causes it to spiral outward from its place of origin close to Earth. If the Moon had formed orbiting in the opposite direction, this effect would have caused it to spiral inward to a catastrophic collision. Such a dramatic fate awaits Triton, Neptune's large and backward-orbiting moon.*

tween Earth and Moon to increase. Besides measurement by laser, the recession of the Moon can also be detected in the fossil record. Daily and annual layers in Devonian horn corals show that about 400 million years ago there were 400 days in each year, the Moon was closer, and Earth was spinning faster. The coupling of these two effects is due to conservation of angular momentum, the same physical law that allows ice skaters to spin faster by pulling their arms against their bodies. The outward movement of the Moon would be reversed if the Moon happened to be orbiting in the opposite direction. Instead of retreating, it would approach Earth and would eventually collide with it. Although we have nothing to fear from Earth's moon in this respect,

Triton, the large moon of Neptune, is in a retrograde orbit and will collide with Neptune within a few hundred million years.

A New Account of the Moon's Origin

A quite remarkable aspect of the Moon is that its formation appears to have been highly unlikely, a rare chance happening. The origin of the Moon has inspired endless speculation ever since people first gazed into the sky, but interest in this subject peaked in 1969, when Apollo 11 landed on the Moon and returned lunar rocks to earthbound laboratories. A major goal of the feverish research activity on these samples was to determine how the Moon, the "Rosetta Stone of the solar system," had formed.

Before the Apollo rocks were brought back, the most popular notion was that the Moon had formed cold and consequently would retain records of the earliest history of the solar system. When the lunar samples were returned, there was breathless anticipation that the samples would finally show how the Moon formed. However, no one satisfactorily solved the mystery of the Moon's origin in 1969 or even in the next decade. Ironically, the assembly of a well-agreed-upon theory for the lunar origin was slowed by the fabulous wealth of highly detailed data from the samples. Extensive work did show that the Moon had a violent, high-temperature history and was thus not a benevolent body for preserving records of its earliest past, as had been hoped. The rocks did, however, record exquisite details of the Moon's history between 3 and 4 billion years ago, a time interval when Earth's history is very poorly known.

During the Apollo program, everyone talked about the origin of the Moon; in the following years, though, most lunar scientists worked on details, and the "big picture" was little discussed. As happens at times in science, this situation radically changed when a sort of cosmic convergence occurred at a meeting on the origin of the Moon held at Kona, Hawaii, in 1984. At this meeting, many details of the analysis of lunar samples and advances in theory came to light, and many scientists left Kona convinced that the Moon had a quite peculiar and improbable origin. Theories on the lunar origin usually fit into three categories: it formed in place, it formed elsewhere and then was captured, or it was somehow ejected from Earth. The new idea was, in a sense, a combination of all three (see Figure 10.3). The model was that during its formation, Earth was hit by a Mars-sized (half the diameter of Earth) projectile. Debris from this collision was ejected into space, and some of it remained in orbit, where self-collisions would cause it to form a thin, orbiting ring of rocks analogous to the rings of Saturn. The moon would form from this ring by collision and sticking, the processes of accretion that built most of the bodies in the solar system.

Several important aspects of this process were consistent with the properties of the Moon deduced from lunar sample studies. One was that the great violence of the event would deplete the Moon of the so-called volatile elements. Compared to meteorites, the moon is depleted in elements such as zinc, cadmium, and tin. These relatively volatile elements would have been vaporized in the impact, and the resulting vapor would have had difficulty completely recondensing from hot gas. Stuck in the gas phase, the volatile elements would have been swept into space and lost to the Earth–Moon system. Included in the lost elements and compounds were nitrogen, carbon, and water. One of the surprising initial findings of Apollo was that the lunar samples were exceedingly dry. Unlike terrestrial rocks, they contained no detectable water. Another remarkable property of lunar samples is that they are highly depleted in the siderophile, or iron-loving, elements that tend to concentrate in the metallic iron cores of planets. When planetary cores form by the molten iron sinking to the planet's center, the siderophile elements (such as platinum, gold, and iridium) are incorporated into the falling iron and are highly depleted from the crustal and mantle materials left above. That the lunar rocks were depleted in siderophile elements was unexpected, because the Moon cannot have a substantial iron core. The mean density of the Moon is 3.4 times the density of water, very similar to that of rocks on the lunar surface and much lower than that of Earth (5.5 times the mean density of water). If it had a substantial core of dense metallic iron, the mean density of the Moon would be higher than is observed. Seismic and magnetic data also show no evidence of a significant core.

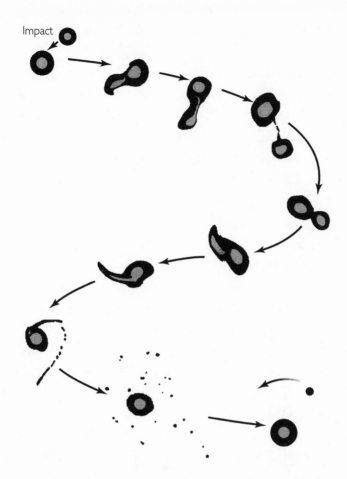

Impact

Figure 10.3 *Impact origin of the Moon as modeled by Cameron and Canup (1998). A body several times more massive than Mars impacts the edge of the half-grown Earth with spectacular effects. After a glancing blow, the two distorted bodies separate and then recombine. The metallic cores (light gray) of both bodies coalesce to form Earth's core, while portions of the mantles (black) of both bodies are ejected into orbit and accumulate to form the Moon. After its formation the Moon spiraled outward, a process that continues to the present time. To produce such a massive moon, the impacting body had to be the right size, it had to impact the right point on Earth, and the impact had to have occurred at just the right time in the Earth's growth process.*

The collision model solves the siderophile mystery by suggesting that metallic cores formed in both Earth and the projectile before the collision. In the collision, both cores ended up in the center of Earth, and the debris ejected into orbit was mainly from the mantles of both bodies. This sequestering of siderophiles explains why gold and platinum are so rare on the Moon and in the crustal rocks of Earth. The impact ejection of mantle materials from both the giant impactor and the target Earth is consistent with some of the remarkable similarities between the trace element content of the Moon and that of rocks from Earth's mantle. It is also consistent with Earth and the Moon being identical in isotopic composition.

A collisional origin is very attractive, but did it actually happen? In earlier times, it was often imagined that the accretion of planets occurred in a regular fashion by collision with small bodies. This was why the Moon was thought to have formed cold. A body that grows by accretion of small objects does not bury heat in its interior. Even though the bodies may collide at high velocity, if they are small they produce small craters, and the energy of impact can be largely radiated back into space. For an impact to eject enough material to form the Moon, the colliding body has to be huge, a Mars-size body. Theoretical modeling by George Wetherill, a Medal of Science–winning planetary scientist at the Carnegie Institute of Washington Department of Terrestrial Magnetism, showed that a natural consequence of the accretion process is that several large bodies do form in each planet's accretion zone. The growth process includes the impact of several bodies each of which carries more than 10% of the mass of the final planet.

In the case of Earth, these big bodies were the size of Mars and larger. Their impact not only ejected material into space to form the Moon but also injected considerable amounts of heat into Earth's mantle. This heat input and great violence led to the forging of Earth's core during the accretion phase, before the planet was fully formed. This is in contrast with a planet that formed cold, by slow accretion of small bodies. Core formation requires high internal temperatures so that blobs of molten iron descend through the mantle to reach the core. Such a planet could form a core only after long-term buildup of radioactive heat from the decay of uranium, potassium, and thorium. In Earth's case, the early heat from accretion of large bodies led to core formation as accretion occurred. The Moon's formation occurred after its core formed. Both bodies had differentiated and already had metal cores at the time of collision.

Recent computer simulations by A. Cameron and colleagues provide the best fit with the actual properties of Earth and the Moon when the Moon-forming collision occurs when Earth has only grown to about half of its final mass and the impactor mass is about a fourth of the final mass of Earth. When the collision occurred, the effects on both bodies were incredible. They briefly fused, but then inertial effects on the resulting plastic mass literally ripped the assembly into two major pieces. The fragments separated for several hours but then fell back in response to the forces of gravity. After a few violent oscillations, the planet finally settled down. Like droplets ejected from a stone tossed into a pond, a small amount of material was ejected and formed a debris ring around Earth. Materials in the ring were derived from the silicate mantles of both bodies. The Moon formed from that disk by accretion of solid particles over a period of a few tens of thousands of years.

When it finally formed, the Moon was only 15,000 miles from the planet's surface. With the Moon so close, Earth would have been spinning at such a rate that the day would be only 5 hours long. The height of the tides would have been fantastic, and as noted earlier, the resulting heat probably melted the surface of the planet. The importance of this heat may be purely academic, because Earth would still have been exceedingly hot just from the great impact. The impact event was so energetic that it actually vaporized rocks, forming a "silicate atmosphere" that survived for a short time before it cooled and condensed as a silicate rain. All of these effects would have been detrimental to any attempts by life to establish an early presence on the planet.

Although this is somewhat of a conjecture, it is possible that Earth's violent early history seeded the eventual development of plate tectonics. The large-scale heating would have led to formation of a magma ocean covering the surface of the planet. Differentiation of this "ocean" may have spawned the first rocks that could form long-term continents. This violent early history must also have had severe effects on the ocean and atmosphere.

If the Earth's formation could be replayed 100 times, how many times would it have such a large moon? If the great impactor had resulted in a retrograde orbit, it would have decayed. It has been suggested that this may have happened for Venus and may explain that planet's slow rotation and lack of any moon. If the great impact had occurred at a later stage in Earth's formation, the higher mass and gravity of the planet would not have allowed enough mass to be ejected to form a large moon. If the impact had occurred earlier, much of the debris would have been lost to space, and the resulting moon would have been too small to stabilize the obliquity of Earth's spin axis. If the giant impact had not occurred at all, the Earth might have retained a much higher inventory of water, carbon, and nitrogen, perhaps leading to a Runaway Greenhouse atmosphere.

Of the many elements of the Rare Earth Hypothesis, the presence of our huge Moon seems to be one of the most important and yet most perplexing. Without the large Moon, Earth would have had a very unstable atmosphere, and it seems most unlikely that life could have progressed as successfully as it has. Even with Earth's relatively stable long-term climate, it still took over 90% of the planet's lifetime to date for land animals to develop. Unfortunately, there is no evidence on how common large moons are for warm terrestrial planets close to their parent stars. We just don't know, and we probably won't for some time. Detection of terrestrial planets will be possible in the coming decades, but detecting their satellites will be much more difficult.

The Moon, our closest neighbor in space, has thus figured prominently in the origin and evolution of life on Earth. Other solar system bodies, though far more distant, have also had effects remarkable enough to suggest that the conditions that promoted development of life on Earth are rare if not unique. A fascinating case in point is a body over 500 million miles from us, the planet Jupiter.

JUPITER

Even in a small backyard telescope, Jupiter is quite a sight. In the eyepiece, it is a disk noticeably flattened because of its high spin rate of two revolutions per day. With parallel equatorial bands and whitish color, its appearance is totally distinct from what Carl Sagan called our "pale blue dot" Earth, a blue planet shrouded with wispy clouds. The most remarkable property of Jupiter visible with a backyard telescope is the presence of four moons, pinpricks of light that can be seen to move over a few hours' time. The rhythmic mathematical dance of the four largest moons, first observed by Galileo in 1612, was a stunning scientific finding, because it resembled a miniature Copernican solar system where orbital motion could be directly observed.

What cannot be seen by telescope, of course, is the exotic interior of this planet. Jupiter is a giant gas ball that gets hotter and denser with depth, but like the other giant planets in the solar system, it has no surface. Jupiter is mostly hydrogen and helium, and deep in its interior, the pressure is so high that electrons are not bound to individual hydrogen atoms but instead move freely from atom to atom, as in a metal. At pressures of a million atmospheres in the interior of Jupiter, hydrogen is actually in a metallic state.

It is enchanting in a small telescope, it has fascinating properties and a rich history, but observed with the unaided eye, this giant planet is just a spot of light—another "star" in the sky. From a distance of half a billion miles, it is difficult to imagine (at least for an astrology agnostic) that this distant planet with its frigid upper atmosphere could have any effect on Earthlings. Remarkably, however, Jupiter's existence and its time and place of formation have profoundly influenced our Earth's ability to provide and maintain a stable environment for life.

A Giant's Influence When the Planets Formed

Jupiter is ten times larger than Earth (and over 300 times its mass), and it is by far the most massive planet orbiting the sun. The origins of Jupiter and its

235

neighbor Saturn (the "Jovian" planets) differ from those of other solar system bodies in that these planets grew largely by direct accretion of gas from the solar nebula in addition to accretion of solids. They are a more democratic sampling of the nebula, and accordingly, they have elemental compositions that are fairly similar to that of the sun, mostly hydrogen and helium. Jupiter formed very quickly, and its rapid growth had major effects on the interior planets, especially those that were trying to accumulate just inside its orbit. For example, a terrestrial planet was in the process of forming about halfway between Jupiter and the sun, but because Jupiter formed first, its development was aborted. This failed planet is now the asteroid belt, a region where some of the original planetesimals and their fragments still survive but were never assembled into a real planet. The largest of these planetesimals is the asteroid Ceres, a rounded object 1000 kilometers in diameter.

The asteroids are the source of meteorites, and detailed examination of these ancient rocks provides deep insight into the nature of planet formation. Most meteorites are ancient "rubble pile" mixtures of materials as old as the solar system. They are the oldest radiometrically dated rocks. The meteorites indicate that initially there was a period of growth when colliding materials led to accretion but that later, during most of the solar system's history, collisions have occurred with such high energy that they have led to erosion and disruption, not growth.

The effects that aborted the formation of a planet in the asteroid belt also severely affected the formation of Mars. Mars is often described as the most Earth-like planet, but in fact it is only half the size and one-tenth the mass of Earth. Presumably, both Mars and the asteroid planet would have grown to the size of Earth if the rapid growth of their giant neighbor had not occurred. If this had been the case, the solar system might have ended up with three truly Earth-like planets, each with oceans and advanced life forms living on or near their surfaces. If Mars had been as large as Earth, it probably would have retained a denser atmosphere, and, with the increased radioactive heat that comes with additional mass, it is likely that it would have been more volcanically active, perhaps driving plate tectonics. (Because of its smaller size, the volcanic activity on Mars is only a few percent of what occurs on Earth.) A larger Mars would also have had a larger core, presumably producing a larger magnetic field. One of the most critical shortcomings of the planet Mars, from the point of view of hospitality to life, is that it almost entirely lacks a global magnetic field. Thus electrically charged particles (the solar wind) flowing outward from the sun played a major role in sputtering the Martian atmosphere off into space. A substantial, Earth-like magnetic field deflects the solar wind and protects the atmosphere from erosion.

If Earth had been a little closer to Jupiter, or if Jupiter had had a somewhat larger mass, then the "Jupiter effect" that aborted the formation of the asteroid planet and nearly ruined the formation of Mars could also have affected Earth, rendering it a smaller planet. And if Earth had been smaller, its atmosphere, hydrosphere, and long-term suitability for life would surely have been less than ideal.

Following the discovery that Martian meteorites arrive at Earth at a rate of half a dozen each year, some investigators have suggested that Mars played a role in seeding Earth with life. The reasoning is that Mars is tougher than Earth to sterilize globally. Ironically, this aspect of habitability is caused by the lack of a Martian ocean. During the first half-billion years of the history of the solar system, during what is called "the period of heavy bombardment," the terrestrial planets were hit by projectiles larger than 100 kilometers in diameter. On Earth, impacts of such magnitude vaporized part of the ocean, and heat from the impact and the resulting greenhouse effect could warm the entire surface of the planet to sterilization temperature. On Mars, with no ocean, such an impact could cause great regional damage but would not sterilize the whole planet. With its thin atmosphere, surface heating would also be more rapidly radiated into space. The low total abundance of water on Mars may thus be the result of these giant impacts, coupled with the planet's lower mass and surface gravity. If the early Mars did have oceans, they may have been effectively ejected into space by impacts. Even if there were more water on the young Mars, most of the early impact history of Mars occurred on a planet that was dry compared to Earth. If life evolved on both planets, it may have been destroyed on Earth, once or even several times, while it survived on Mars.

There are sound reasons to believe that life may have formed during a limited window of opportunity. This window of time may have closed before the end of heavy bombardment on Earth. It is thus possible that the present life on Earth is of Martian origin, transported to Earth by meteorites ejected by major impacts. If Mars had been Earth-like, with oceans, then it too would have been sterilized by impact. If Mars had been larger and had had a denser atmosphere, it would also have been much more difficult for impacts to eject meteorites into space.

A Distant Sentinel

Jupiter also played a crucial role in purging the inner solar system of bodies left over from planet formation. Jupiter is 318 times more massive than Earth, and it exerts enormous gravitational influence. Its gravitational interactions very efficiently scatter bodies that approach it, and it has largely cleaned out stray bodies from a large volume of the solar system. In the early solar system, there were tremendous numbers of small bodies that had escaped incorporation into planets, but over half a billion years, most of the larger ones inside the orbit of Saturn disappeared. They were accreted by planets, ejected out of the solar system, or incorporated into the Oort cloud of comets. Jupiter was the major cause of this purging of the middle region of the solar system.

The objects that still impact Earth today are planetesimals that managed to survive in three special ecological niches: the Oort comet cloud beyond Pluto, the Kuiper belt of comets just beyond the outer planets, and the asteroid belt, that special refuge located between Mars and Jupiter. The current impact rate averages one 10-kilometer body every 100 million years. The impact of just such a body occurred 65 million years ago, the time of the K/T extinction that ended the age of the dinosaurs. George Wetherill of the Carnegie Institute of Washington has estimated that the flux of these 10-kilometer bodies hitting Earth might be 10,000 times higher if Jupiter had not come into being and purged many of the leftover bodies of the middle region of the solar system. If Earth had been subject to collisions with extinction-causing projectiles every 10,000 years instead of every 100 million years, and fairly frequently with even larger bodies, it seems unlikely that animal life would have survived.

Do most planetary systems harbor planets like Jupiter? Ours has two (Jupiter and Saturn), and the detection of Jupiter-mass planets around other stars suggests that Jupiters exist in other planetary systems, but their frequency is still unknown. It is likely that many planetary systems do not have Jovian planets. The standard formation model of Jupiter has it accreting a large, solid core before it can begin accreting gas, its main constituent. The necessary conditions (available mass and rapid accretion before gas is lost from the planetary system) may not be common, and so Jupiter-mass planets may be rare in other planetary systems. When planetary systems lack a Jovian planet to guard the outer boundary of the terrestrial planet region, the inner planets may not be capable of supporting more than microbial life.

The Origin—and Fortuitous Stability—of Jupiter

Why and how did Jupiter form where it did? It is generally believed that Jupiter's formation began with the accretion of a solid core. This core grew by collision and sticking of dust, ice, rocks, and larger bodies—a process similar to the accretion of Earth. Jupiter, however, formed outside the "snow line," a special place in the solar system where water vapor condensed to form ice grains and the presence of "snow" in this region would enhance the density of solid matter and accelerate the accretion process. The mystery is why the proto-Jupiter grew so rapidly. Apparently, Jupiter grew to a mass of 15 Earths before Mars grew to 10% of an Earth mass. David Stevenson at Cal Tech has suggested that outward migration of water vapor and condensation at the "snow line" may have provided larger concentrations of condensed matter at this location, thus speeding up the formation of the embryonic Jupiter.

Jupiter's growth to a giant planet began when the rock–ice core mass reached 15 Earth masses. At this mass, the gravity of the core can pull in and hold hydrogen and helium, the light gases that account for 99% of the mass of the nebula. When this gas accretion process begins, it is very dramatic because the rate of accretion of gas is proportional to the square of the mass already accreted. In other words, the bigger it gets, the faster it grows. If gas could be continually fed to it, it would gobble up the Universe in a relatively short time! What actually happens is that Jovian planet formation depletes its feeding zone of matter, which in turn truncates planet formation. And although the general properties of this process might be modeled, it just seems to have been by chance that our Jupiter formed as it did.

Because it cleans our solar system of dangerous Earth orbit–crossing asteroids and comets, Jupiter has had a beneficial influence on life on Earth. However, it appears that we have been quite lucky that the Jupiter in our solar system has maintained a stable orbit around the sun. A Jupiter and a giant neighbor like Saturn are a potentially deadly couple that can lead to disastrous situations where a planetary system can literally be torn apart. Recently, it has become possible to use powerful computers to determine the stability of the orbits of Jupiter and Saturn over the lifetime of the planetary system. There are minor chaotic changes but no major changes, and the solar system, at least to a first approximation, is stable over its lifetime. However, this would not be case if either Jupiter or Saturn were more massive or if the two were closer together. It would also be dangerous to have a third Jupiter-sized planet in a planetary system. In an unstable system the results can be catastrophic. Gravitational perturbations among the planets can radically change orbits, make them noncircular, and actually lead to the loss of planets ejected into interstellar space. It is possible for chaotic disruption to occur even after a system has been stable for billions of years, and in the worst cases, planets can be spun out of the planetary system, escaping the gravitational hold of the star. A life-bearing planet ejected into galactic space would have no external heat source to warm its surface and no sunlight to provide energy for photosynthesis. Although instability might start with just two planets, the effects would spread to them all. In less severe cases, the orbits of the planets would become highly elliptical, and the changing distance between planets and the central star would prevent the persistence of conditions required for stable atmospheres, oceans, and complex life.

Numerical calculations first indicated that some planetary systems might become unstable, and recent observations provide evidence that this actually does occur. At the present time, planets are being discovered around other stars by detecting small velocity changes—Doppler shifts—of the central star. Of the planets that have been detected, many are Jupiter-mass planets far from the star, with highly noncircular orbits. This is quite different from the solar system, where all giant planets have quite circular orbits. It is generally agreed that the best explanation for the elliptical orbits is that these are planets whose orbits have been altered by other planets, possibly by ejection of another planet into interstellar space.

It is during the formation of planetary systems that Jupiter-like planets pose the most draconian threat to terrestrial planets. With radial order similar to that in the solar system, the terrestrial planets in the habitable zone of typical planetary systems would form close to the star, the Jovian planets farther out. There is reason to believe that this is the "natural way," because the formation of Jovian planets probably occurs only in the colder, more distant regions outside the "snow line," mentioned previously. It is also to be expected that giant gaseous planets like Jupiter could not form close to a star because tidal forces (the differential force of gravity) would disrupt a planet in its diffuse, early stages. When a proto-Jupiter was very diffuse and too close to the star, the differential force of gravity between the sides near the star and far from it would pull the forming planet apart. It was quite surprising, then, that several of the first extrasolar planets discovered were found to be Jupiter-mass planets very close to their central stars—closer than Mercury to the sun. All of these "hot Jupiters" have highly circular orbits, and it is difficult to imagine that they actually formed in these locations.

A popular explanation of this phenomenon is that the giant planets in these systems actually did form at Jupiter-like distances but that their orbits decayed and the planets spiraled inward. This cannot happen in an evolved planetary system, but it can in the early, solar nebula phases when extensive gas and dust still exist in the regions between the planets. Doug Lin at the University of California at Santa Cruz has calculated that spiral waves generated

241

in the solar nebula phase can extract energy from a young Jupiter and cause its orbit to spiral inward. In many cases the planet actually hits the star; in others the inward drift stops before collision occurs. The observed giant planets that are very close to stars may be examples of this inward drift. Events such as this can be calamitous for terrestrial planets. When a Jupiter spirals inward, the inner planets precede it and are pushed into the star. If our Jupiter had done this, Earth would have been vaporized long before life-tolerant conditions were ever established on its surface. Lin has suggested that out solar system may have had several Jupiters that actually did spiral into the sun, only to be replaced with a newly formed planet. Perhaps Jupiter is at its "right" distance from the sun only because it was the last one to form and it formed at a time when the solar nebula had weakened to the point where orbital decay ceased to be important.

The programs to detect extrasolar planets have revealed that nearly all of the planets found either are "hot Jupiters" in circular orbits close to the star or describe elliptical orbits farther from the star. All of these are "bad" Jupiters whose actions and effects should preclude the possibility of these systems having animal life on Earth-like planets in the habitable zones of the parent stars. These life-unfriendly planetary systems have been found around 5% of the nearby stars. The search techniques are most effective for detecting Jupiters close to stars, however, and at present they cannot detect Jupiter-mass planets at our Jupiter's distance from the sun, nor can they detect planets less massive than Jupiter. They could not presently detect our solar system from the distance of nearby stars, and so it is possible that up to 95% of the nearby solar-like stars have "regular" planetary systems similar to our own, with terrestrial planets close to the star and Jovian planets in circular orbits farther out. On the other hand, it appears that most other "Jupiters" so far detected would have prevented the development of animal life anywhere in their respective solar systems.

The Moon and Jupiter are two factors causing us to believe that complex life requires disparate influences. We shall see, in the next chapter, how we might put this hypothesis to the test.

Testing
the Rare Earth
Hypothesis

The Rare Earth Hypothesis is the unproven supposition that although microscopic, sludge-like organisms might be relatively common in planetary systems, the evolution and long-term survival of larger, more complex, and even intelligent organisms are very rare. The observations on which this hypothesis is based are as follows: (1) Microbial life existed as soon as Earth's environment made it possible, and this nearly invincible form of life flourished over most of Earth history, populating a broad range of hostile terrestrial environments. (2) The existence of larger and more complex life occurred only late in Earth history, it occurred only in restricted environments, and the evolution and survival of this more fragile variant of terrestrial life seem to require a highly fortuitous set of circumstances that could not be expected to exist commonly on other planets. This hypothesis can be tested.

Throughout human history, people have wondered what lies beyond the limits of the known world. This instinctive obsession has driven humans (and perhaps other species) to expand their own territories. This haunting question permeates mythology and religion and has provoked some of the deepest of human thoughts. In earlier times, the phrase *beyond the known world* may have referred to regions only hundreds to thousands of miles distant. In modern times, these musings extend to actual worlds—to other planets. Over the past century and a half, the great advances in science and in our understanding of nature and physical processes have refined our ability to imagine other worlds realistically and evaluate the possibilities of life beyond Earth. We now actually have the knowledge and technological tools to begin the serious search for alien life, and for the first time in history, we have the capability to test the Rare Earth Hypothesis. The tests that can be done are of two types. One consists of efforts to detect the presence of microbial life in other bodies in the solar system. The discovery of living microorganisms or fossil evidence of microorganisms would support the contention that microbial life originates readily, that it forms frequently, and that we might expect it to occur in numerous bodies that have warm, wet environments somewhere in their interiors. The search for microbial life can be done in the solar system by sending specialized probes to seek life directly with *in situ* analysis techniques.

The second test of the Rare Earth Hypothesis is the search for evidence of advanced life forms, which might range from simple, multicellular organisms to large animals. We see no evidence of advanced life in the solar system, except on Earth, so the main search for advanced life will focus on planetary systems around nearby stars. These searches will be conducted via large, space-borne telescopes. Both the *in situ* detection of microbial life and the telescopic detection of advanced life are in the planning stages, and both have high priority with funding agencies in the United States and Europe. This is a very exciting time. It is our first opportunity to actually study the processes that lead to the origin, evolution, and survival of life in the Universe.

ADVANCED LIFE

An alien astronomer, viewing Earth from a great distance, could detect the presence of life on the planet with comparative ease. This could not be done

244

directly, by imaging, but rather indirectly, by spectral analysis of the composition of the atmosphere. Even with the most grandiose alien telescopes, it is doubtful that extraterrestrial astronomers could directly detect organisms, groups of organisms, or even the most immense structures life has formed: coral reefs, forests, forest fires, red tides, city lights, the Great Wall of China, freeways, and dams. At best, images of Earth would only slightly resolve what Carl Sagan referred to as the "pale blue dot." The principal clue alerting distant astronomers to the presence of life would be a spectacular and unmistakable signature. Spectral analysis of infrared light would reveal that life plays such a major role on the planet that it controls the composition of the atmosphere.

Spectrum of a Life-Bearing Planet

Earth's atmosphere would actually be quite "unnatural" for a nonbiotic planet. It is clearly different from the nearly pure carbon dioxide atmospheres of its neighbors, Mars and Venus. The mix of nitrogen, oxygen, and water vapor is chemically unstable and would never arise on a dead planet. Without life, nitrogen and oxygen in the presence of water would combine to form nitric acid and become a dilute acidic component of the ocean. Earth's peculiar atmosphere is not in chemical equilibrium, and it succeeds in disobeying natural chemical laws only because of the presence of life. The most peculiar aspect of the atmosphere is the abundance of free oxygen. Oxygen is the most abundant element in the whole Earth (45% by weight and 85% by volume!), but in the atmosphere, it is a highly reactive gas that would exist only at trace levels in the atmosphere of a terrestrial planet devoid of life. Oxygen is a poisonous gas that oxidizes organic and inorganic materials on a planetary surface; it is quite lethal to organisms that have not evolved protection against it. The source of atmospheric oxygen is photosynthesis, the miraculous biological process that utilizes the energy of sunlight to convert carbon dioxide to pure oxygen and organic material. Ironically, it was the long-term photosynthetic production of this poisonous gas, and life's adaptation to it, that made complex and energetic life possible on Earth. Except for the noble gas argon, all of the major atmospheric constituents are also processed and recycled on short time scales via biological processes.

Our distant alien astronomers would realize that life exists on Earth as soon as they detected, in its infrared spectra, absorption bands due to the presence of carbon dioxide, ozone, and water vapor (see Figure 11.1). Nitrogen and normal oxygen (O_2) are the major atmospheric gases, but they do not produce detectable absorption effects. The telltale bands arise because of the way Earth interacts with sunlight. Earth's surface is warmed by visible sunlight, and it reradiates in the infrared. The incoming energy is in the visible region of the spectrum (near 0.5 micrometer in wavelength) where the sun, with its surface temperature of 5400°C, emits most of its energy. The atmosphere is largely transparent to visible wavelengths, and the light that is not reflected into space is largely absorbed by Earth's surface. This energy heats the surface to "room temperature"—only about 5% of the absolute temperature of the surface of the sun. Earth's surface cools itself to balance exactly (averaged over time) the absorption of sunlight by radiating infrared radiation back into space. Because of Earth's relatively cooler temperature, the bulk of this energy is in the "thermal infrared" spectral region near 10 micrometers in wavelength. Atmospheric transmission of parts of this spectral region are blocked because of absorption by certain gases. Water vapor, ozone, and carbon dioxide absorb part of the outgoing infrared radiation and block its escape from Earth. This process and these same gaseous species are the root cause of the atmospheric greenhouse effect that prevents Earth's oceans from freezing. All of these "greenhouse" molecules are minor constituents, but they cause warming effects of some 40°C above the temperature of an atmosphere that is totally transparent in the infrared. They also provide a very strong spectral signature to be seen by alien astronomers. Water makes up a few percent of the atmosphere, CO_2 is currently only 375 parts per million, and ozone occurs only in the parts-per-billion range. Although rare, they absorb significant chunks of the infrared radiation streaming into space. The outgoing infrared has significant absorption dips at wavelengths of 7, 10, and 15 micrometers, respectively, for water, ozone, and carbon dioxide.

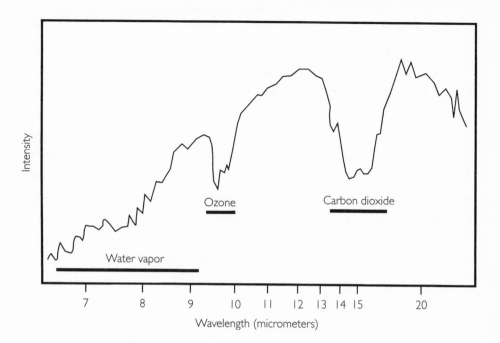

Figure 11.1   *The hypothetical infrared spectrum of an Earth-like planet with life. The abundances of water vapor, carbon dioxide, and ozone would be clues indicating that the planet was in the habitable zone of its star and that life was producing oxygen.*

Seeking Life's Spectral Signature

Observation of ozone, CO_2, and H_2O in the atmosphere of an unknown planet strongly suggests habitable conditions and the presence of life. Water vapor at moderate levels can indicate that a planet is essentially in the habitable zone of a star. Water vapor abundance depends on the temperature of an atmosphere as well as on the availability of surface water. CO_2 abundance also provides a clue to the "habitability" of a terrestrial planet. At least on Earth, the ability to keep this dangerous greenhouse gas locked in sediments requires moderate surface temperatures and an active land–ocean weathering cycle, whereby "excess" CO_2 can be sequestered in carbonates and thus kept out of the atmosphere. The strengths of the CO_2 and H_2O absorptions provide strong clues for identifying planets that are truly Earth-like, with land, oceans, and moderate surface temperatures. The detection of ozone signals the presence of active, photosynthetic life. Ozone (O_3) is a molecule composed of three oxygen atoms. It is a highly chemically active molecule, so it is not very stable. It is produced in the atmosphere by the interaction of ultraviolet light from the sun and normal oxygen (O_2). Light splits O_2 into individual atoms, and they in turn react with O_2 to form ozone. Only a minuscule fraction of the atmosphere's oxygen is in the form of ozone, but it provides strong infrared absorption. Ozone implies the presence of oxygen, which, in sufficient concentration, implies the existence of life. Maintaining moderate oxygen levels requires continuous production to balance the many processes that lead to its removal from the atmosphere.

Several space-borne projects have been proposed to detect infrared spectral signatures that would suggest Earth-like planets, oceans, moderate surface temperatures, and the presence of biological activity. A project in the planning stage at NASA is the Terrestrial Planet Finder, and a similar project being studied by the European Space Agency (ESA) is appropriately called Darwin. Both of these missions would use very large telescopes to image terrestrial planets and take infrared spectra. The basic challenge is to measure the spectral signature of terrestrial planets close enough to their parent stars to be in the habitable zones. This is a formidable task, in part because of the faintness of planets but also because of the proximity of these planets to the star. Viewed from a great distance, Earth itself would look very close to the sun and comparatively very faint. From the distance of the nearest stars, the angular separation of Earth and the sun is comparable to the diameter of a quarter as viewed from a distance of 4 miles. This is near the limit of what can be resolved with conventional ground-based telescopes, but it is easily within the reach of space telescopes and of ground-based telescopes with adaptive optics to reduce the blurring effects of the atmosphere.

The major problem with the study of extrasolar planets, even for nearby stars, is that planets are much fainter than stars. At visual wavelengths, Earth is seen only by the sunlight it reflects. The sun is a billion times brighter than the reflected "Earth-light," and in any telescope system, the glare from such an intensely bright object overwhelms the image of a faint nearby planet. In the infrared, at wavelengths of 10 micrometers, the situation is better. The sun is much fainter, and this is the peak of the region where the warm Earth radiates energy back to space. At this wavelength the sun is only ten thousand times brighter than Earth, and separation of the two images is feasible via special techniques involving interferometry.

The grand plan for detecting life in extrasolar planets involves the construction of very large telescopes. It is not efficient—or perhaps even possible—to build a single-mirror telescope large enough to detect individual extrasolar planets, and this limitation suggests using groups of smaller telescopes combined to work in concert. The ability to resolve ultrafine angular detail depends on the diameter of the telescope. Telescopes together in an array have resolving power equivalent to a single mirror as large as the entire array. Present plans for the Terrestrial Planet Finder call for a cluster of four telescopes, each with an individual mirror 4 to 8 meters in diameter. These could be mounted on a truss or could be free flyers with a total separation of about 100 meters. In either case, the separation of the individual telescopes would have to be controlled to incredible precision: small fractions of the wavelength of light. The telescopes would combine light beams and would function as an interferometer with a very special property. Their sensitivity would be a minimum in the center of the imaged field and a maximum at a slight offset equal to the expected angular spacing of a planet and the star. When the telescopes were then pointed directly at the star, the image of the star would effectively be attenuated by a factor of a million, whereas the light from a nearby planet would not be attenuated.

This special design produces a "null" on the star, minimizing the great difference in brightness between planet and star. The technique uses interference, the same process the produces the iridescence of soap bubbles and the brilliant color of some butterfly wings. A laser pointer projected onto a distant wall shows an analogous effect in reverse. It has a bright spot in the middle that is surrounded by faint dark and bright rings. The planet finders will use interference to produce the opposite effect: a null in the center and sensitivity in rings around it. Doing precision interferometry with such huge telescopes in space will require extraordinary effort and billions of dollars. The technique is just at the edge of available technology but appears practicable. The system could search for Earth-like planets around several hundred of the nearest stars, and in it lies our best hope for detecting life outside the solar system any time soon.

Seeking Intelligent Life

Another approach to the search for extrasolar life is the detection of radio signals sent by other civilizations. The modest attempts to pick up any such signals have generated much interest, speculation, and debate. It is true that the most powerful radio telescopes on Earth could receive signals from similar telescopes, aimed directly at Earth, from any other spot in the galaxy. Considerable thought has also been devoted to what wavelengths would be used for communication and what types of information might be sent. The laws of physics and radio propagation in the galaxy suggest that the best wavelengths are in what is known as the "water hole" near 20 centimeters. This activity is commonly referred to as SETI, the Search for ExtraTerrestrial Intelligence. In 1990, NASA funded a modest SETI effort, but its budget was cut after only a few years, before the program got seriously under way. Senator Proxmire gave SETI one of his famed Golden Fleece awards and fumed "not a penny for SETI." Others in influential positions were concerned about ridicule of a program that might run for decades or even millennia seeking faint radio signals from other civilizations, and public funding for SETI searches has been very limited. (Just as it is a problem on Earth, funding would probably be a critical factor on other worlds too. On Earth, in our most economically prosperous times, we cannot even listen. Sending the signals would be much more complex and costly.)

Unfortunately, it is very difficult to know if SETI is an effective use of resources. If the Rare Earth Hypothesis is correct, then it clearly is a futile effort. If life is common *and* it commonly leads to the evolution of intelligent creatures that have long, prosperous planetary tenures, then it is possible that enlightened aliens might be beaming signals off into space. A key factor in deciding whether SETI makes sense involves the lifetime of civilizations with radio technology. Does such a civilization last only centuries before nuclear war, starvation, or some other calamity causes its decline? Or does it last forever? In the most optimistic minds, "Star Trek" societies might populate the stars. But even if they do, it is a real question whether any of them would or could beam enormous amounts of radio power into space to potential audiences that are prevented by the vast interstellar distances from ever returning the message in a timely manner. There probably are other civilizations in the galaxy that have radio telescopes, but the vast numbers of stars and the vast distances involved are barriers that may always keep SETI more an experiment of the imagination than a large-scale scientific endeavor. An exception might be made for the limited number of nearby stars that have planetary systems. If some of these are found to have Earth-like planets with atmospheric compositions indicative of life, then the public might support either sending signals or listening. And, of course, even though we do not intentionally beam radio messages to nearby stars, Earth is a potent transmitter of radio power emanating from radar, television stations, and other sources.

MICROBIAL LIFE IN THE SOLAR SYSTEM

The search for microbial life in the solar system began in earnest with Apollo 11. Although it was clear that the Moon was not a teeming abode of life, it was thought that the Moon might provide clues to early life or at least to prebiotic chemical conditions. The astronauts and the samples they collected underwent elaborate quarantine, lest lunar microbes attack Earthlings like the disastrous diseases carried across the Atlantic Ocean just 450 years earlier. Before Apollo, some thought that the Moon was similar in composition to primitive meteorites—that it might contain abundant carbon and water in the form of hydrated minerals. A popular theory for the Moon's origin was that it formed elsewhere and was captured during a close encounter with Earth. This theory had it that the Moon initially had a much smaller orbit and then retreated outward in response to tidal effects. Harold Urey, the Nobel Prize–winning chemist (and one of the leading pioneers in the field of planetary science), imagined that the Moon would have passed so close to Earth that the immense tidal interaction would have resulted in parts of the ocean sloshing into space and landing on the lunar surface. Although few believed that any living organisms could flourish in the harsh environment of the airless Moon, Urey thought the Moon might retain critical records of prebiotic chemistry and desiccated remains of the earliest forms of life on Earth. Urey called the Moon the Rosetta Stone of the solar system.

When the Apollo 11 samples were returned, the first tests were toxicological, to see whether the samples had any dire effects on terrestrial life. Some of the priceless cargo of lunar soil was fed to rats and placed in the root systems of growing plants. No negative effects were observed, and detailed analysis of the rocks and soils revealed no organic material of biological origin. There was carbon, but it all appeared to have been derived from impacting meteoritic bodies and implantation by the solar wind. As mentioned above, the lunar samples were extraordinarily "dry"; they contained no bound water. The Moon was found to be a lifeless body that did not even contain the building blocks of life or a life-supporting environment.

The Viking program was the only space mission that directly included life detection among its goals. This extraordinary program involved four spacecraft: two that landed on the surface of Mars for detailed *in situ* studies and two that went into orbit for global-scale mapping and to relay lander information to Earth. With the possible exception of the Hubble Space Telescope, Viking was the most expensive NASA mission launched purely for scientific exploration. (Apollo had a large scientific component, but the mission was largely motivated by national priorities—getting to the moon first.) The Viking missions cost about 4 billion 1999 dollars and required robotic spacecraft to land on another planet to conduct chemical searches for the presence of life. The first mission landed on Mars in 1976, the American Bicentennial, and many of the scientists involved referred to each other as the Vikings of '76. The large, cowboy-style brass belt buckles that of them many wore, with the mission logo engraved thereon, are still seen at various meetings of planetary scientists and engineers.

Viking was an enormously difficult and successful mission. And yet, in a sense, the Viking mission was a failure in that it did not detect life. Not only did it not detect life, but the results revealed the Martian surface to be a highly inhospitable environment for life. There was less carbon in the soil of Mars than there was on the Moon, and worse, the presence of highly oxidizing conditions indicated that organic material could not survive in the soil. If a dead mouse were buried in shallow Martian soil, its carbon would be converted to carbon dioxide, which would flow into the atmosphere. The results from Viking drove many nails into the coffin of the belief that Mars was an Earth analog that might harbor life.

The Viking missions carried three major life detection experiments. Each was a miniature, highly specialized chemical laboratory designed to detect chemical changes characteristic of biological activity. Each lander had a retractable arm with a scoop at the end. One of the joys of the mid-1970s was watching these scoops actually dig trenches and collect samples on Mars, the famous red planet of so many science fiction stories. The scoops would dig soil samples and drop them through a screen into the analysis instruments. The major life detection experiments were the gas exchange (GEX), labeled release (LR), and pyrolitic release (PR) techniques. The first data, 8 days after the Viking I landing, came from the GEX experiment, and the results were positive—or appeared to be so. A gram of soil was placed in the experimental chamber, and a small amount of water and nutrient was added. Two days later a large amount of newly generated oxygen gas was detected, an expected signal of biological activity. The LR experiment also yielded positive results only a day later. In the labeled release experiment, water and nutrient labeled with radioactive carbon-14 was added to a soil sample, and the equipment recorded whether C^{14}-labeled carbon dioxide or methane was released.

The signal was again positive, and—startlingly more positive, in fact, than in many soils on Earth! In the pyrolitic experiment no nutrient or water was added, but the soil was exposed to C^{14}-labeled carbon dioxide and to carbon monoxide gas and light. After an exposure, the soil was heated (pyrolized) to see whether C^{14}-labeled material would be released from any newly formed organic compounds. There was a weak but positive signal.

In spite of high hopes and expectations, however, the Viking scientists had built in some backup tests. The instruments were designed so that multiple experiments and samples could be run repeatedly, as in a "real" Earth-based laboratory. Repeated tests showed that the "positive" results could be attributed to unusual chemical properties of Martian soil. With no ozone layer to block it, the harsh ultraviolet light from the sun lands directly on the soil and produces highly oxidizing and reactive compounds, such as peroxides, that can produce the reactions observed. After severe heating that would have killed any terrestrial organism, the soils still yielded "positive signals." The Viking team's interpretation of their data indicate that instead of the actions of living Martian organisms, the observed results were caused by surface chemical reactions of nonbiological origin.

The Viking landers did not convincingly detect Martian life, but they did show how difficult it is to identify microbial life, with unknown properties, on a planet that is quite different from Earth. Viking was capable of detecting organisms in most of Earth's surface materials, but our planet teems with life, and a gram of typical soil contains over a billion individual organisms. Viking found that the surface soils on Mars did not and could not support any of the forms of life found on Earth. If life does exist on Mars, we must look for it in subterranean regions beneath the frozen "cryosphere" at depths where liquid water can persist. Future missions cannot simply look for life in spoonfuls of surface soil; they must search the warm, wet regions beneath the inhospitable surface. To directly reach wet rock, future searches for living organisms will have to drill. And drilling must not be done just anywhere, because the frozen cryosphere normally extends to depths of several kilometers. Instead of attempting to drill so deeply, future missions will search out rare geothermal hot spots where liquid water might reach close to the surface. These missions will search the "Yellowstones" of Mars. It is also possible that samples of life-bearing rocks might be found on the surface as debris excavated by impact craters. Studies of terrestrial and lunar craters have shown that large rocks—some the size of houses—can be lifted from considerable depth and deposited on the crater rim. Organisms could not live in these cold dry rocks, but they might survive in a dormant state for thousands or even millions of years.

Although finding living creatures that could be observed to reproduce would be the most convincing discovery, the next searches for life on Mars will have less ambitious goals. They will search for fossils or for chemical, isotopic, or mineralogical indicators of past life. Even if Mars is currently a totally sterile planet, it may well have harbored life in its distant past. Channels and other surface features suggest that Mars was much more Earth-like three or four billion years ago. Mars occasionally had liquid water on its surface, and it probably had lakes or larger bodies of water persisting for moderate periods of time beneath thick ice crusts. If the Rare Earth Hypothesis is correct and life forms readily, then we would expect life to have evolved on Mars during the early period of its evolution when it had more Earth-like surface conditions. The search for life on Mars is a key test of this hypothesis.

Searching for microscopic fossils or other indicators of life is enormously complex and is difficult to do in a convincing way with spacecraft instruments. Space instruments must be designed to perform very specific tasks. Constraints on power, mass, cost, reliability, and remote operation in a hostile environment mean that the kitchen sink and most of the other items that scientists would like to include on the spacecraft are inevitably left behind. Space instruments are usually marvels of their time, but their capabilities rarely are competitive with those of their heavy, power-guzzling and inelegant siblings that are used for day-to-day work in Earth-bound laboratories. The most important limitation of spacecraft instruments is their inflexibility to adapt to new findings. They usually do what they are designed to do, but not more. This differs considerably from normal laboratory studies, where initial results provide new insights and lead to investigations that were not previously anticipated. For these reasons, the most detailed searches for life and fossils on Mars will require sample return. Searches for evidence of life in Martian meteorites have produced intriguing results and have shown what types of investigations could be done on returned samples. The first such mission is now planned for launch in 2005. Although they pose an enormous technical challenge, the Martian sample return missions offer the best current hope for finding evidence of life on Mars. Once returned to Earth, the samples will be examined with the most sensitive instruments we have for clues of the past existence of microbial organisms. Even if life exists on Mars, of course, its discovery may require a series of exploration and sample return missions.

In addition to Mars, there are many other bodies in our solar system that might have microbial life. These include the three outermost large moons of Jupiter (Europa, Ganymede, and Callisto), the large moon of Saturn (Titan), and possibly other moons as well. Europa presently is the most attractive prospect apart from Mars. Images of its surface show a complex landscape of shifting ice and mysterious ridges. Heated by tidal energy, liquid water lurks beneath the surface. Lesser amounts of water or brine are also thought to exist inside Callisto and Ganymede. Although twice as far away, Titan is also an exploration target of great interest. Its dense nitrogen atmosphere and hydrocarbon-rich surface is tantalizing—in spite of cryogenic temperatures at its surface. In 2004 the Cassini mission will parachute an instrumented package onto the surface of Titan. It is not designed to detect life, but this probe will measure environmental parameters that are important to life.

Most of our discussion up to this point has been couched in relatively qualitative terms. It's time now to look at some efforts that have been made to quantify the probability that life will evolve and persist, and to suggest a few numbers of our own. This is the subject of the next chapter.

Assessing
the Odds

Indeed, the only truly serious questions are ones that even a
child can formulate. Only the most naïve of questions are
truly serious.

—Milan Kundera, *The Unbearable Lightness of Being*

"Do you feel lucky? Well do ya?"

—Clint Eastwood, *Dirty Harry*

How rare is Earth? We have arrived at the end of a long grocery list of ingredients seemingly necessary to make a planet teeming with complex life. It involves material, time, and chance events. In this chapter we will try to assess these various factors and their relative importance; all can be thought of as probabilities. In some cases we understand these probabilities, but in others almost no research has been done, and our questions, like those referred to in the quotation above, are the simple questions of children—questions as yet with no answers. Some of these questions can thus be tackled only with our imagination. Others will be answered by the space voyages and instrumented investigations we discussed in the previous chapter.

Let us begin by imagining we have the power to observe 100 solar nebulae coalesce into stars and the planets that will encircle them. How many of these events will yield an Earth-like planet with animal life?

As we have seen, the first step in preparing the way for a habitable environment is the formation of a suitable star: one that will burn long enough to let evolution work its wonders, one that does not pulse or rapidly change its energy output, one without too much ultraviolet radiation, and most important, perhaps, one that is large enough. Of the 100 applicants, perhaps only two to five will yield a star as large as our sun. The vast majority of stars in the Universe are smaller than our sun, and although smaller stars could have planets with life, most would be so dim that Earth-like planets would have to orbit very close to their star to receive energy sufficient to melt water. But being close enough to get adequate energy from a small star leads to another problem: tidal lock, the condition where the same side of the planet always faces the sun. A tidally locked planet is probably unsuitable for animal life.

What if we increased the number to 1000 planetary systems, so that we might expect 20 stars of our sun's size or greater to be born? Even these numbers are too small to yield a high probability that we will find a truly Earth-like planet. Perhaps a better way to envision the various odds is to re-create the scenario that led to the formation of our solar system and then run through the process once again in a thought experiment. Stephen Jay Gould used this type of mental reconstruction in his interpretation of the Cambrian Explosion. In his 1989 book *Wonderful Life*, Gould described the exercise as follows:

> I call this experiment "replaying the tape." You press the rewind button and, making sure you thoroughly erase everything that actually happened, go back to any time and place in the past—say, to the seas of the Burgess Shale. Then let the tape run again and see if the repetition looks at all like the original.

A THOUGHT EXPERIMENT

In our case, we will replay the tape of our planet's formation. We begin with a planetary nebula of exactly the mass and elemental composition that created our solar system. According to most theorists, this might create a star identical to our sun—but then it might not. For instance, the spin rate of the new star might be different from that of our sun, with unknown consequences. Well, then, 1000 such solar nebulae, *perhaps* 1000 clones of dear old Sol. Not so, however, with the planets coalescing out of this mix. If we rerun this particular tape, we will in all probability not get a repeat of our solar system with its nine planets, its one failed planet (now the asteroid belt), a Jupiter and three other gas giants orbiting outside four terrestrial planets; and a halo of comets surrounding the entire mix. Now we enter the realm of multiple contingencies. Of the 1000 newly formed planetary systems, none is likely to be identical to our solar system today—just as no two people are identical. In a coalescing planetary system many processes, including planetary formation, may be chaotic.

Planets form in what are known as "feeding zones," regions where various elements come together and eventually coalesce into planetesimals, which finally aggregate into a planet. Recent work by planetary scientists shows that the spacing of planets will probably be fairly regular. There might be as few as six planets or as many as ten or even more. James Kasting of Penn State University believes that planetary spacing is not accidental—that the positions of planets are highly regulated, and that if the solar system were to re-form many times, we would get the same number of planets each time. Yet the observational evidence to date does not back up the theory. The extrasolar planets that have been discovered exhibit an enormous diversity of spacing and orbits; their positions are not nearly so orderly as the theory suggests they should be. Ross Taylor, an astronomer who received the prestigious Leonard Award in 1998, disputes Kasting's views. "Clearly," he maintains, "the conditions that existed to make our system of planets are not easily reproduced. Although the processes of forming planets around stars are probably broadly similar, the devil is in the details."

No one knows whether a planet the size of Jupiter would always form or whether there would be a couple of planets like Mars instead. A planet would probably form in about the position of Earth, but it could be larger or smaller, somewhat closer to the sun or father away. Would the material (physical) quantities be essentially the same? Would plate tectonics develop? Would there be the same amount of water—and would that water end up on the surface of the planet, rather than locked up in its mantle or lost to space? Would there be few threats to life from Earth orbit–crossing asteroids? What is the chance that our Moon would form again, if, as we believe, it is important in making Earth a stable place conducive to animal diversification?

Even if all of these events occurred more or less the way they have, would life form again? And given life, would animal life appear once more? Can there be animal life without the utterly chance events that occurred in Earth's history, such as a Snowball Earth or an inertial interchange event, for instance?

Let us reorganize (and rephrase) this set of questions in the following way. We might ask: How many of all planets in the Universe are terrestrial planets (as opposed to the gas giants such as Jupiter, for instance)? What is the percentage for all planets of the Universe? (In our solar system there are five, but if we add the larger moons, that number more than triples.) Of the terrestrial planets in the Universe, how many have enough water to form an ocean (either as water or as ice)? Of those planets with oceans, how many have any land? Of those with land, how many have continents (rather than, say, scattered islands)? Yet these questions are only for the infinitesimally small slice of time we call the present. All of these conditions are subject to change.

BUYING TIME: THE PERSISTENCE OF OCEANS AND MODERATE TEMPERATURE

As we have tried to show in the preceding chapters, the most important lesson from Earth's history is that it takes time to make animals—long periods of environmental stability with global temperatures staying at about half the boiling point of water or less. Hence we need to add the time component to each question. For instance, what percentage of planets that have oceans keep them for a billion years, or 4 billion years, or 10 billion years, for that matter?

Of all the factors important in assessing the odds of once again getting (or finding) a world with animal life, one factor stands paramount: water. Earth succeeded in acquiring its ark-load of animals and complex plants—and then *keeping* them—for more than half a billion years (so far) because it retained its oceans for more than 4 billion years. Moreover, if our analysis of the sedimentary record is correct, for the last 2 billion years it maintained the oceans at average temperatures less than 50°C. Also—at least for the last 2 billion years—the oceans have been maintained at a chemical composition conducive to the existence of complex animal life: at a salinity and pH favorable to the formation and maintenance of proteins. The oceans are clearly the cradle of animal life—not fresh water, not the land, but the saltwater oceans have spawned every animal phylum, every basic body plan that exists or has existed on our planet.

Discovering how Earth acquired its supply of water is one of the most critical concerns of the new field of astrobiology. As we pointed out in an earlier chapter, water was not abundant in the inner regions of the solar system when the planets formed. There was far more water in the outer regions of the solar system than among the inner planets. Where did our water come from?

Although where our oceanic water came from is still the subject of debate, everyone agrees that it must have arrived during planetary accretion, with perhaps significant volumes added during the period of heavy bombardment. Ironically, the volume of water eventually found on Earth may be related to the formation of Earth's core. When the iron- and nickel-rich core formed, most of the water found in the coalescing planet was consumed in oxidation processes whereby oxygen bound up in water was used to make iron and nickel oxides. It is the residual water that makes up the oceans. Perhaps that residual quantity was significantly enhanced by water carried by comets after Earth's initial formation, perhaps not. In either case, the oceans reached approximately their present volume by 3.8 billion years ago. But this does not mean they were at their present area. Don Lowe of Stanford University has estimated that before 3 billion years ago, less than 5% of the surface was land. Until about 2.5 billion years ago, the chemistry of this world-girding ocean was controlled largely by interactions with the oceanic crust beneath it and with Earth's mantle, whose by-products interacted in the oceans at mid-ocean ridges and rifting areas. It is estimated that because this early Earth was much warmer than the Earth we know, the area of this zone of ocean-mantle contact was as much as six times that found today.

Earth's atmosphere was also very different from that of today. There was no oxygen, and there was a great deal more carbon dioxide—perhaps 100 to 1000 times as much as today. Earth's surface temperature was higher than it is now because more heat was emanating from the interior and because of the warming generated by the extensive CO_2 and other greenhouse gases in the atmosphere. Earth's internal generation of heat was an important factor; the sun at this time was much fainter, and delivering perhaps a third less energy, than at the present time.

What would have happened if Earth had stayed a water world? Probably global temperatures would have remained high or even increased. For animal life to form, the temperature had to drop from the levels acknowledged to have been characteristic of Archaean time. A drop in global temperature while the sun was getting hotter required a drastic reduction of atmospheric CO_2—a reduction of the greenhouse effect. Thus some means of removing CO_2 had to be brought to bear. As we saw in Chapter 9, the most effective way to do this is through the formation of limestone, which uses CO_2 as one of its building blocks and thus scrubs it from the atmosphere. But significant volumes of limestone form today only in shallow water; the most effective limestone formation occurs in depths of less than 20 feet. In deeper water, high concentrations of dissolved CO_2 slows or inhibits the chemical reactions that lead to limestone formation. There is evidence of deep-water, inorganic limestone formation in very old rocks on Earth, as demonstrated by John Grotzinger and his team from M.I.T. These studies showed that the early Earth's ocean may have been saturated in the compounds that can produce limestone and thus could have precipitated limestone in deeper water at that time, removing carbon dioxide from the atmosphere as a consequence. However, Grotzinger points out that occurrences of carbonate rocks during the Early Archaean—roughly the first billion years of Earth's existence—are rare. And this is only partly due to the rarity of rocks of this age. It looks as though the central mode of removing carbon dioxide from the atmosphere—the formation of carbonate rocks—seldom occurred.

To form limestone in significant volumes, then, shallow water is needed, but on a planet without continents, shallow water is in short supply. If the volume of water on a planet is low enough that significant areas of shallow water are available even without continents, there is no problem. On Earth and other planets with significantly deep oceans, however, without continents the shallow-water regions would not be large enough for the necessary limestone formation. Thus when planets have too much water on their surfaces—when their oceans are too deep—there is no natural brake on carbon dioxide buildup. Water temperatures will rise as planetary temperatures rise.

What about underwater weathering? James Kasting has pointed out to us that an all-water world can indeed regulate its temperature. He rightly notes that as oceanic water temperature rises, it eventually causes weathering of limestone on the bottom of the sea. Although much less efficient than the weathering of continental material, this mechanism will indeed produce a feedback mechanism. Yet to heat water temperatures sufficiently to serve as a global thermostat a planet may well exceed the critical 40°C mark that is the upper temperature limit of animal life. If plate tectonics on Earth had not created increasingly large land areas (and, as a by-product of that, massive areas next to the continents with shallow-water regions where limestone could easily form), Earth might well have reached global temperatures greater than animal life could tolerate. And had global temperatures exceeded 100°C, the oceans would have boiled away, the gigantic volumes of water becoming steam in the atmosphere. This would have spelled a catastrophic ending of all life on the surface of planet Earth.

The removal of carbon dioxide is called CO_2 drawdown. On Earth, it was accomplished because of continent formation, which took place during a relativity brief interval of Earth history. From perhaps 2.7 to 2.5 billion years ago there occurred a rapid buildup of continental areas. This buildup resulted in the land surface increasing from perhaps 5% to about 30%. This marked change had equally profound effects on the atmosphere–ocean system.

With the formation of continents as a result of plate tectonics, ocean chemistry became dominated by weathering by-products of the continents. As continents weather, or undergo the chemical and mechanical break-down of rock material, river runoff carries enormous volumes of these chemicals into the sea, where they can greatly affect ocean chemistry and cause mineralization—such as carbonate formation. Larger continents also, paradoxically enough, meant larger shallow-water regions, for the emergence of continents created the shallow continental shelves as well as large inland seas and lakes. Thus the following sequence unfolded: Large shallow regions were created; nutrient influx from continental regions increased; the amount of plant material on Earth (mainly in the surface regions of the shallow seas and on shallow sea bottoms) skyrocketed; and oxygen production began in earnest. All of these events opened the pathway to the eventual evolution of animals.

The critical question is why, on Earth, the volume of water was sufficiently large to buffer global temperatures, but small enough so that shallow seas could be formed by the uplifting of continents. If Earth's ocean volume had been greater, even the formation of continents would not have produced shallow seas. To show that there can be great relative volumes of oceans on a planet, we need only look at Jupiter's moon Europa, where the planet-covering ocean (now frozen) is 100 kilometers thick. No Mt. Everest rising from the sea floor would ever poke through an ocean even half that deep. There would be none of the shallows necessary for limestone formation and no continental weathering.

What about the situation where the oceans are *lower* in volume than they were on Earth? If the continents covered two-thirds of Earth's surface (rather than their present day one-third), would we have animal life? The great mass extinction of the late Permian almost ended animal life because of high temperatures. With greater continental area, we might expect temperature swings to have been even greater, and the prospects for continued existence of at least land animals far lower, because large land areas create very high and very low seasonal temperatures. Large land areas also reduce CO_2 drawdown, because carbonate formation takes place almost exclusively in oceans. On land-dominated worlds, opportunities for life to thrive would thus be reduced.

It appears that Earth got it just right. Without continents there seems a strong likelihood that a planet will become too hot (especially because main-sequence stars such as the sun increase their energy output through time, and planets cannot move away from this increasing heat source). With too much continental area, the opposite is likely to happen, as continental weathering draws down carbon dioxide so much that glaciations ensue. Earth may have been headed down the path toward a global mean temperature so high as to boil away its oceans, or perhaps still cool enough to retain its oceans but yet too warm for complex metazoans to evolve. Animals are not thermophiles.

How much land area is "just right," and how much is too little or too much? The answer probably depends on the given planet's distance from the sun. A planet whose orbit dictates that it receives less energy from its star than Earth does from the sun might need a greater amount of ocean cover (assuming that an increased sea surface creates warmer planetary temperatures because of greater greenhouse effects from CO_2 buildup).

The relative areas of land and sea affect more than just planetary temperature. If plate tectonics is not operating, there will be no continents, only large numbers of seamounts and islands (whose number will be dictated by the amount of volcanicity, itself a function of a planet's heat flow). And without continents, a planet's ocean may never achieve a chemistry suitable for animal life. Sherwood Chang of NASA cites an example of this. In 1994 Chang proposed that without substantial weathering (which can occur only when there is substantial land area to weather), the early ocean of an Earth-like planet would remain acidic—a poor environment for the development of animal life. It seems that water worlds might be quite fecund habitats for short periods of time but might not achieve the long-term temperature or chemical stability conducive to animal life.

The Importance—and Sheer Chance Occurrence—of Our Large Moon

Although many scientists have been doggedly pursuing the various attributes necessary for a habitable planet—Michael Hart, George Weatherill, Chris McKay, Norman Sleep, Kevin Zahnlee, David Schwartzman, Christopher Chyba, Carl Sagan, and David Des Marais come to mind—one name stands out in the scientific literature: James Kasting of Penn State University.

Kasting notes that whether habitable planets exist around other stars "depends on whether other planets exist, where they form, how big they are, and how they are spaced." Kasting stresses, as we do, the importance of plate tectonics in creating and maintaining habitable planets, and he suggests that the presence of plate tectonics on any planet can be attributed to the planet's composition and position in its solar system. But one of Kasting's most intriguing comments is related to our Moon. Kasting notes that the obliquity (the angle of the axis of spin of a planet) of three of the four "terrestrial" planets of our solar system—Mercury, Venus, and Mars—has varied chaotically.

> Earth is the exception, but only because it has a large moon. . . . If calculations about the obliquity changes in the absence of the moon are correct, Earth's obliquity would vary chaotically from 0 to 85 degrees on a time scale of tens of millions of years were it not for the presence of the Moon. . . . Earth's climatic stability is dependent to a large extent on the existence of the Moon. The Moon is now generally believed to have formed as a consequence of a glancing collision with a Mars-sized body during the later stages of the Earth's formation. If such moon-forming collisions are rare . . . habitable planets might be equally rare.

We have accumulated a laundry list of potentially low-probability events or conditions necessary for animal life: not only Earth's position in the "habitable zone" of its solar system (and of its galaxy), but many others as well, including a large moon, plate tectonics, Jupiter in the wings, a magnetic field, and the many events that led up to the evolution of the first animal. Let us explore what these conditions might mean for life beyond Earth.

The Odds of Animal Life Elsewhere, and of Intelligence

In the 1950s, astronomer Frank Drake developed a thought-provoking equation to predict how many civilizations might exist in our galaxy. The point of the exercise was to estimate the likelihood of our detecting radio signals sent from other technologically advanced civilizations. This was the beginning of sporadic attempts by Earthlings to detect intelligent life on other planets. Now called the Drake Equation in its creator's honor, it has had enormous influence in a (perhaps necessarily) qualitative field. The Drake Equation is simply a string of factors that, when multiplied together, give an estimate of the number of intelligent civilizations, N, in the Milky Way galaxy.

As originally postulated, the Drake Equation is.

$$N^* \times fs \times fp \times ne \times fi \times fc \times fl = N$$

where:

N^* = stars in the Milky Way galaxy
fs = fraction of sun-like stars
fp = fraction of stars with planets
ne = planets in a star's habitable zone
fi = fraction of habitable planets where life does arise
fc = fraction of planets inhabited by intelligent beings
fl = percentage of a lifetime of a planet that is marked by the presence of a communicative civilization

Our ability to assign probable values to these terms varies enormously. When Drake first published his famous equation, there were great uncertainties in most of the factors. There did (and does) exist a good estimate for the number of stars in our galaxy (between 200 and 300 million). The number of star systems with planets, however, was very poorly known in Drake's time. Although many astronomers believed that planets were common, there was no theory that *proved* star formation should include the creation of planets, and many believed that the formation of planetary systems was exceedingly rare. During the 1970s and later, however, it was assumed that planets were common; in fact, Carl Sagan estimated that an average of ten planets would be found around *each star*. Even though no extra solar planets were found until the 1990s, their discovery seemed to vindicate those who believed planets were common. But is it so? A new look at this problem suggests that planets may indeed be quite rare—and thus the presence of animal life rarer still.

Are Stars with Planets Anomalous?

We now know that planetary formation outside our own system does indeed occur. The recent and spectacular discovery of extrasolar planets, one of the great triumphs of astronomical research in the 1990s, has proved what has long been assumed: that other stars have planets. But at what frequency? It may be that a substantial fraction of stars have planetary systems. To date, however, astronomers have succeeded in detecting only giant, "Jupiter-like" planets; available techniques cannot yet identify the smaller, rocky, terrestrial worlds. Now that numerous stars have been examined, it appears that *only about 5% to 6% of examined stars have detectable planets*. Because only large gas-giant planets can be detected, this figure really shows that Jupiter clones close to stars or in elliptical orbit are rare. But perhaps it indicates that planets as a whole are rare as well.

The evidence that planets may be rare comes not so much from the direct-observation approach of the planet finders (such as the Marcy/Butler group) but from spectroscopic studies of stars that appear similar to our own sun. The studies of those stars around which planets have been discovered have yielded an intriguing finding: They, like our sun, are rich in metals. According to astronomers conducting these studies, there seems to be a causal link between high metal content in a star and the presence of planets. Our own star is metal-rich. In a study of 174 stars, astronomer G. Gonzalez discovered that the sun was among the highest in metal content. It appears that we orbit a rare sun.

Other new studies also require us to question the belief that planetary systems such as our own are common. At a large meeting of astronomers held in Texas in early 1999, it was announced that 17 nearby stars had been observed to be orbited by planets the size of Jupiter. Astronomers at the meeting were also puzzled by an emerging pattern: None of the extrasolar planetary systems resembles the sun's family of planets. Geoff Marcy, the world's leading planet finder, noted that "for the first time, we have enough extrasolar planets out there to do some comparative study. We are realizing that most of the Jupiter-like objects far from their stars tool around in elliptical orbits, not circular orbits, which are the rule in our solar system." All of the Jupiter-sized objects either were found in orbits much closer to their sun than Jupiter is to our sun, or, if they occurred at a greater distance from their sun, had highly elliptical orbits (observed in 9 of the 17 so far detected). In such planetary systems, the possibility of Earth-like planets existing in stable orbits is low. A Jupiter close to its sun will have destroyed the inner rocky planets. A Jupiter with an elliptical or decaying orbit will have disrupted planetary orbits sunward, causing smaller planets either to spiral into their sun or to be ejected into the cold grave of interstellar space.

It is still impossible to observe smaller, rocky planets orbiting other stars. Perhaps such planets—which we believe are necessary for animal life—are quite common. But perhaps this is a moot point. We have hypothesized that animal life cannot long exist on a planet unless there is a giant, Jupiter-like planet within the same planetary system—and orbiting outside the rocky planets—to protect against comet impacts. It may be that Jupiters like our own, in regular orbits, are rare as well. To date, all tend to be in orbital positions that would be lethal, rather than beneficial, to any smaller, rocky planets.

PLANET FREQUENCY AND THE DRAKE EQUATION

All predictions concerning the frequency of life in the Universe inherently assume that planets are common. But what if the conclusions suggested by emerging studies—that Earth-like planets are rare, and planets with metal rarer still—are true?

This finding has enormous significance for the final answer to the Drake Equation. Any factor in the equation that is close to zero yields a near-zero final answer, because all the factors are multiplied together. Carl Sagan, in 1974, estimated that the average number of planets around each star is ten. Goldsmith and Owen, in their 1992 *The Search for Life in the Universe,* also estimated ten planets per star. But the new findings suggest greater caution. Perhaps planetary formation is much less common than these authors have speculated.

To estimate the frequency of intelligent life, the Drake Equation hinges on the abundance of Earth-like planets around sun-like stars. The most common stars in the galaxy are M stars, fainter than the sun and nearly 100 times more numerous than solar-mass stars. These stars can generally be ruled out because their "habitable zones," where surface temperatures could be conducive to life, are uninhabitable for other reasons. To be appropriately warmed by these fainter stars, planets must be so close to the star that tidal effects from the star force them into synchronous rotation. One side of the planet always faces the star, and on the permanently dark side, the ground reaches such low temperatures that the atmosphere freezes out. Stars much more massive than the sun have stable lifetimes of only a few billion years, which might be too short for the development of advanced life and evolution of an ideal atmosphere. As we noted earlier, each planetary system around a 1-solar-mass star will have *space* for at least one terrestrial planet in its habitable zone. But will there actually *be* an Earth-sized planet orbiting its star in that space? When we take into account factors such as the abundance of planets and the location and lifetime of the habitable zone, the Drake Equation suggests that only between 1% and 0.001% percent of all stars might have planets with habitats similar to those on

Earth. But many now believe that even these small numbers are overestimated. On a universal viewpoint, the existence of a galactic habitable zone vastly reduces them.

Such percentages seem very small, but considering the vastness of the Universe, applying them to the immense numbers of stars within it can still result in very large estimates. Carl Sagan and others have mulled these various figures over and over. *They ultimately arrived at an estimate of one million civilizations of creatures capable of interstellar communication existing in the Milky Way galaxy at this time.* How realistic is this estimate?

If microbial life forms readily, then millions to hundreds of millions of planets in the galaxy have the *potential* for developing advanced life. (We expect that a much higher number will have microbial life.) However, if the advancement to animal-like life requires continental drift, the presence of a large moon, and many of the other rare Earth factors discussed in this book, then it is likely that advanced life is very rare and that Carl Sagan's estimate of a million communicating civilizations is greatly exaggerated. If only one in 1000 Earth-like planets in a habitable zone really evolves as Earth did, then perhaps only a few thousand have advanced life. Although it could be argued that this is too pessimistic, it may also be much too optimistic. Even so, we cannot rule out the possibility that Earth is not unique in the galaxy as an abode of life that has just recently developed primitive technologies for space travel and interplanetary radio communication.

Perhaps we can suggest a new equation, which we can call the "Rare Earth Equation," tabulated for our galaxy:

$$N^* \times fp \times ne \times fi \times fc \times fl = N$$

where:

N^* = stars in the Milky Way galaxy
fp = fraction of stars with planets
ne = planets in a star's habitable zone
fi = fraction of habitable planets where life does arise

fc = fraction of planets with life where complex metazoans arise

fl = percentage of a lifetime of a planet that is marked by the presence of complex metazoans

And what if some of the more exotic aspects of Earth's history are required, such as plate tectonics, a large moon, and a critically low number of mass extinctions? When any term of the equation approaches zero, so too does the final result. We will return to this at the end of this chapter.

If animal life is so rare, then intelligent animal life must be rarer still. How can we define intelligence? Our favorite definition comes from Christopher McKay of NASA, an astronomer, who defines intelligence as the "ability to construct a radio telescope." Although a chemist might define intelligence as the ability to build a test tube, or an English professor as the ability to write a sonnet, let us for the moment accept McKay's definition and follow the lines of reasoning he sets out in his wonderful essay "Time for Intelligence on Other Planets," published in 1996. Much of the following discussion comes from that source.

McKay points out that if we accept the "Principle of Mediocrity" (also known as the Copernican Principle) that Earth is quite typical and common, it follows that "intelligence has a very high probability of emerging but only after 3.5 billion years of evolution." This supposition is based on a reading of Earth's geological record, which suggests to most authors that evolution has undergone a "steady progressive development of ever more complex and sophisticated forms leading ultimately to human intelligence." Yet McKay notes—as we have tried to emphasize in this book—that evolution on Earth has *not* proceeded in this fashion but rather has been affected by chance events, such as the mass extinctions and continental configurations produced by continental drift. Furthermore, we believe that not only events on Earth, but also the chance fashion in which the solar system was produced, with its characteristic number of planets and planetary positions, may have had a great influence on the history of life here.

McKay breaks down the critical events in the evolution of intelligence on Earth as shown in the accompanying table.

Event	When It Happened on Earth (millions of years ago)	How Long It Took to Complete (millions of years)	Possible Minimum Time (millions of years)
Origin of life	3800–3500	< 500	10
Oxygenic photo-synthesis	< 3500	< 500	Negligible
Oxygen environments	2500	1000	100
Tissue multicellularity	550	2000	Negligible
Development of animals	510	5	5
Land ecosystems	400	100	5
Animal intelligence	250	150	5
Human intelligence	3	3	3

We can certainly quibble with some (or all) of his numbers, especially his estimate of when life first arose on Earth, for we think it occurred far earlier than 3800 to 3500 million years ago. Yet these estimates are probably not off by orders of magnitude. McKay's point is that complex life—and even intelligence—could conceivably arise faster than it did on Earth. If we accept McKay's figures, a planet could go from an abiotic state to the home of a civilization building radio telescopes in 100 million years, as compared to the nearly 4 billion years it took on Earth. But McKay also concedes that there may be other factors that require a long period:

> What is not known is whether there is some aspect of the biogeo-chemical processes on a habitable planet—for example, those dealing with the burial of organic material, the maintenance of habitable temperatures as the stellar luminosity increases gradually over its main sequence lifetime, or global recycling by tectonics—that mandates the long and protracted development of the oxygen-rich biosphere that occurred on Earth. Other important unknowns include the effect of solar system structure on the origin of life and its subsequent evolution to advanced forms.

273

His inference is that plate tectonics has slowed the rise of oxygen on Earth. But it also may be necessary to ensure a stable oxygenated habitat, just as having the correct types of planets in a solar system is important as well.

In their 1996 essay "Biotically Mediated Surface Cooling and Habitability," Schwartzman and Shore tackle this same problem and reach a different conclusion: They believe that the most critical element in determining the rate at which intelligence can be acquired is a potentially habitable planet's rate of cooling. Their point is that complex life such as animals is extremely temperature-limited, with a very well-defined upper temperature threshold. Although some forms of animal life can exist in temperatures as high as 50°C or sometimes even 60°C, most require lower temperatures, as do the complex plants necessary to underpin animal ecosystems. A maximum temperature of 45°C is probably realistic. It is thus the time necessary for a planet to cool to below this value that is critical, according to these two authors. Many factors affect the time required, including the rate at which a star increases in luminosity through time (which works against cooling), the volcanic outgassing rate (which also works against cooling, because such outgassing puts more greenhouse gases into a planetary atmosphere), the rate at which continental land surface grows (as continents grow, planets usually cool), the weathering rate of land areas, the number of comet or asteroid impacts and their frequency, the size of a star, whether or not plate tectonics exists, the size of the initial planetary oceans, and the history of evolution on the planet.

With this in mind, let us return to our Rare Earth Equation and flesh it out a bit by adding some of the other factors featured in this book.

$$N^* \times fp \times fpm \times ne \times ng \times fi \times fc \times fl \times fm \times fj \times fme = N$$

where:

N^* = stars in the Milky Way galaxy

fp = fraction of stars with planets

fpm = fraction of metal-rich planets

ne = planets in a star's habitable zone

ng = stars in a galactic habitable zone

fi = fraction of habitable planets where life does arise

fc = fraction of planets with life where complex metazoans arise

fl = percentage of a lifetime of a planet that is marked by the presence of complex metazoans

fm = fraction of planets with a large moon

fj = fraction of solar systems with Jupiter-sized planets

fme = fraction of planets with a critically low number of mass extinction events

With our added elements, the number of planets with animal life gets even smaller. We have left out other aspects that may also be implicated: Snowball Earth and the inertial interchange event. Yet perhaps these too are necessary.

Again, *as any term in such an equation approaches zero, so too does the final product.*

How much stock can we put in such a calculation? Clearly, many of these terms are known in only the sketchiest detail. Years from now, after the astrobiology revolution has matured, our understanding of the various factors that have allowed animal life to develop on this planet will be much greater than it is now. Many new factors will be known, and the list of variables involved will undoubtedly be amended. But it is our contention that any strong signal can be perceived even when only sparse data are available. To us, the signal is so strong that even at this time, it appears that Earth indeed may be extraordinarily rare.

Messengers from the Stars

Our planet is not in a special place in the solar system, our
Sun is not in a special place in our galaxy and our galaxy is
not in a special place in the Universe.

—Marcello Gleiser, *The Dancing Universe*

S ome things have to be experienced firsthand; for some wonders, no written description or photograph can substitute—the birth of one's child,
for example, or the first music heard from an actual orchestra, love, sex,
standing before a Monet canvas. One such revelation not so often experienced, is one's first glimpse of the starry night through a telescope.

VIEWING THE UNIVERSE

We have all seen photographs of endless star fields, galaxies, and nebulae. But
no matter how great their beauty, the stars in photographs are lifeless, and no
view of the night sky with the unaided eye, even in the clearest atmosphere,

is like the first view through a small telescope. If looking at the Milky Way with the unaided eye is akin to snorkeling on a coral reef, then adding a telescope is like strapping on scuba tanks: We are no longer tied to the surface but can roam the depths of the star fields and see unimagined splendors amid stars whose numbers are beyond belief. Even with a low-power telescope a new vision emerges; the uncountable pinpricks of light now revealed are seemingly alive, in no way diminished by their passage through corrected lenses. In fact, the stars gain strength, color, and clarity. But the greatest and most lasting impression is the increase in their sheer numbers. The superb double-star cluster in Perseus changes from a dull, unresolvable glow to bountiful diamonds sprinkled on black velvet; the globular cluster in Hercules is transformed from a tiny smudge to scattered grains of light. With time and experience, even greater vistas open up. We discover the joys of other deep-space objects, galaxies and nebulae. And eventually, in the Northern hemisphere, we inevitably find ourselves slowly moving through the crowded star fields of Sagittarius on a dark summer night, the light from this luminous expanse of stars sweeping the senses like a wind, nebulae and galaxies an endless visual melody punctuated by staccatos of brighter suns. Those in the Southern hemisphere witness even more dramatic vistas: the two great Magellenic Clouds looming so close overhead. It becomes spectacle, overwhelming, and ultimately—diminishing. The myriad stars overcome us, so utterly do they trivialize (marginalize? minimize?) our small planet and we who stare out.

The Universe seems to be finite; there are not an infinite number of planets circling the vast number of stars in the ocean of space. But the numbers are immense beyond understanding. We are one of many planets. But as we have tried to show in this book, perhaps not so many as we might hope—and perhaps not so many that we will ever, however long the history of our species, find *any* extraterrestrial animals among the stars surrounding our sun. That is a fate not foreseen by Hollywood—that we may find nothing but bacteria, even on planets orbiting distant stars.

If the Rare Earth Hypothesis is correct—that is, if microbial life is common but animal life is rare—there will be societal implications, or at least some small personal implications. What will be the effect if news comes back from the next Mars mission that there is life on Mars after all—microbial to be sure, but life. Or what if, after astronauts voyage repeatedly to other planets in our solar system, or even to the dozen nearest stars, we find nothing more advanced than a bacterium? What if, at least in this quadrant of the galaxy, we are quite alone, not just as the only intelligent organisms but also as the only animals? How much of our striving to travel into space is the hope of discovering—and perhaps talking to—other *animalia*?

VIEWS OF EARTH THROUGH HUMAN HISTORY

Since the time of the Greeks, science has tried to make sense of the Universe and of our place in it. More than two millennia ago, a Greek named Thales of Miletus, credited by many as the founder of Western philosophy, was among the first to leave a record of his musings about the place of Earth in the cosmos. Thales thought that the cosmos was an organic, living thing, and in that he may not have been far wrong if bacteria or bacteria-like organisms are as common as we believe them to be in the Universe. Thales's student Anaximander was among the first to place Earth at the center of the cosmos, postulating that Earth was a floating cylinder with a series of large wheels with holes in them rotating around it. The Pythagoreans tried to break from this central-Earth motif, proposing that Earth moved in space and was not the center of the Universe. But Earth's centrality was restored by members of Plato's school and became exalted by the students of Aristotle. Eudoxus placed Earth at the center of 27 concentric spheres, each of which rotated around it. Soon two schools of thought competed: the "sun-centered" model of Aristarchus and the Earth-centered model of Ptolemy. The latter held sway through the Middle Ages.

During the Middle Ages, Earth was not only regarded as the center of the Universe but was again believed to be flat. St. Thomas Aquinas made Earth a sphere again but codified its place as the center of the Universe. It was

Nicholas Copernicus who finally shattered the notion of an Earth-centered Universe and put the sun at the center of all orbits. But even with this great leap forward, the sun remained at the center of the Universe as well, according to Copernicus in his revolutionary book of 1514, *Commentariolus*.

Copernicus forever destroyed the myth that our Earth lay at the center of the Universe, with the sun and all other planets and stars revolving around us; his work eventually led to the concept of a "Plurality of Worlds"—the idea that our planet is but one among many. This has now been described as the "Principle of Mediocrity," also known as the Copernican Principle. Yet an even greater blow came with the invention of the telescope. There is still debate about who built the first optical telescope, although Dutch optician Johannes Lippershey obtained the first official license for construction of a telescope in 1608. The device was an immediate sensation, and by 1609 this revolutionary new instrument found its way into the hands of Galileo, who built his own soon after hearing of the concept. Before Galileo, telescopes had been used to assess the terrestrial world (and for various military applications), but Galileo pointed his into the heavens and forever changed our understanding of the cosmos.

Galileo quickly surmised that there are far more stars in the sky than anyone had guessed. He discovered that the Milky Way is made up of uncountable individual stars. He observed the Moon, discovered satellites revolving around Jupiter (and in so doing showed that our Earth could just as conceivably orbit the sun). Earth's central place in the Universe, the fervent belief of Aristotle, was now observationally shown to be wrong. Copernicus had dealt with theory; Galileo and his telescopes dealt with reality. Galileo's message, published in his booklet *Siderius nuncius*, or "Messenger from the Stars," was about the truth told by the stars: that Earth is but one of many cosmic objects. To illustrate his point, he noted the presence of faint patches of light just visible to the unaided eye—objects called nebulae. Even with his primitive and tiny telescopes, Galileo could see these curious objects far better than anyone before. He thought them to be great masses of stars, made indistinct by their very distance.

The decentralization of Earth continued in relentless fashion. In 1755 Immanuel Kant theorized that a rotating gas cloud would flatten into a disk as it contracted under its own gravity. Kant was familiar with the numerous nebulae of the night skies, the faint glowing patches of luminosity scattered through the heavens. All the early astronomers knew of the faint cloud in the constellation of Andromeda. He knew these objects to be one of many distant groups of stars he called "island Universes." But Kant didn't stop there: He theorized that the sun, Earth, and other planets might have formed in this swirling mass of gas. This concept was taken a step further by Pierre-Simon de Liplike, who speculated in detail about how planetary systems might form from nebular origins. He invoked a dynamical mechanism for the formation of stars and their planets. Earth and the solar system became one of many such systems all formed in the same way.

But how far away were these island Universes? Was there only a single galaxy in the Universe, of which our star was part, or were there many? This debate was not resolved until the early twentieth century, a time when gigantic new telescopes were being constructed and outer space was being probed as never before. The conflict came to a head on April 26, 1920, when Harlow Shapley from the Mount Wilson observatory in California and Heber Curtis from the Allegheny Observatory in Pittsburgh met before the members of the National Academy of Sciences, a clash that came to be known as the Great Debate. The debate ended inconclusively, because it was not yet possible to assess the distance of the nebulae. That soon changed, however, thanks to the efforts of astronomer Edwin Hubble. Using a newly constructed, 100-inch reflecting telescope, Hubble was able to make observations that proved conclusively that the island nebulae were not associated with our Milky Way but were far-distant objects. Even the closest, the Andromeda galaxy, was found to be at least 2 million light-years from Earth and similar in shape to our Milky Way galaxy. The debate was over. The Milky Way is one of a vast number of scattered and widely separated galaxies floating in space. We became even more trivialized—now our *galaxy* was but one of many.

281

Two millennia of astronomers and philosophers removed Earth from the center of the Universe and placed us orbiting a sun that is but one of hundreds of billions in a galaxy itself but one among billions in the Universe. And it was not only astronomers who changed the world view. Einstein showed that there is no preferred observer in the Universe, and quantum mechanics told us that chance is king. Charles Darwin and his powerful theory of evolution demoted humans from the crown of creation to a rather new species on an already animal-rich planet, the chance offspring of larger-scale evolutionary and ecological forces. Nothing special. And yet . . .

The great danger to our thesis (that Earth is rare because of its animal life, the factors and history necessary to arrive at this point as a teeming, animal- and plant-rich planet being highly improbable) is that it is a product of our lack of imagination. We assume in this book that animal life will be somehow Earth-like. We take the perhaps jingoistic stance that Earth-life is every-life, that lessons from Earth are not only guides but also *rules*. We assume that DNA is the only way, rather than only one way. Perhaps complex life—which we in this book have defined as animals (and higher plants as well)—is as widely distributed as bacterial life and as variable in its makeup. Perhaps Earth is not rare after all but is simply one variant in a nearly infinite assemblage of planets with life. Yet we do not believe this, for there is so much evidence and inference—as we have tried to show in the preceding pages—that such is not the case.

OUR RARE EARTH

Let us recap why we think Earth is rare. Our planet coalesced out of the debris from previous cosmic events at a position within a galaxy highly appropriate for the eventual evolution of animal life, around a star also highly appropriate—a star rich in metal, a star found in a safe region of a spiral galaxy, a star moving very slowly on its galactic pinwheel. Not in the center of the galaxy, not in a metal-poor galaxy, not in a globular cluster, not near an active gamma ray source, not in a multiple-star system, not even in a binary, or near a pulsar, or near stars too small, too large, or soon to go supernova. We became a planet where global temperatures have allowed liquid water to exist for more than 4 billion years—and for that, our planet had to have a nearly circular orbit at a distance from a star itself emitting a nearly constant energy output for a long period of time. Our planet received a volume of water sufficient to cover most—but not all—of the planetary surface. Asteroids and comets hit us but not excessively so, thanks to the presence of giant gas planets such as Jupiter beyond us. In the time since animals evolved over 600 million years ago, we have not been punched out, although the means of our destruction by catastrophic impact is certainly there. Earth received the right range of building materials—and had the correct amount of internal heat—to allow plate tectonics to work on the planet, shaping the continents required and keeping global temperatures within a narrow range for several billion years. Even as the Sun grew brighter and atmosphere composition changed, the Earth's remarkable thermostatic regulating process successfully kept the surface temperature within livable range. Alone among terrestrial planets we have a large moon, and this single fact, which sets us apart from Mercury, Venus, and Mars, may have been crucial to the rise and continued existence of animal life on Earth. The continued marginalization of Earth and its place in the Universe perhaps should be reassessed. We are not the center of the Universe, and we never will be. But we are not so ordinary as Western science has made us out to be for two millennia. Our global inferiority complex may be unwarranted. What if Earth is extremely rare because of its animals (or, to put it another way, because of its animal habitability)?

The possibility that animal life may be very rare in the Universe also heightens the tragedy of the current rate of extinction on our planet. Earlier, we suggested that the rise of an intelligent species on any planet might be a common source of mass extinction. That certainly seems to be the case on Earth. And if animals are as rare in the Universe as we suspect, it puts species extinction in a whole new light. Are we eliminating species not only from our planet but also from a quadrant of the galaxy as a whole?

To understand the rates of extinction on Earth today, one has only to examine the plight of tropical rainforests. Forests have been a part of this planet for more than 300 million years, and although the nature of species has changed over that long period, the nature of the forests has changed little. The forests are the great Noah's ark of species on this planet. Although the land surface of our globe is only one-third that of the oceans, it appears that 80% to 90% of the total animal and plant biodiversity of the planet inhabits the land, and most of that diversity is found in tropical forests. As we destroy these forests, we destroy species. It has been estimated that between 5 and 30 million species of *animals* live in the tropical rainforests and that only about 5% of these are known to science. The fossil record tells us quite clearly that the world has attained the highest level of biological diversity ever in its history. There are also disturbing and unmistakable signs that this plateau in the number of species on Earth has been crested and the biodiversity of Earth is diminishing.

There appear to be several forces driving a reduction of biodiversity—a *destruction* of biodiversity, to be less delicate. The most important seems to be the rapid increase in human population. Ten thousand years ago there may have been at most 2 to 3 million humans scattered around the globe. There were no cities, no great population centers. There were fewer people on the globe than are now found in virtually any large American city. Two thousand years ago the number had swelled to perhaps 130 to 200 million people. Our first billion was reached in the year 1800. If we take the time of origin of our species as about 100,000 years ago, it seems that it took our species 100,000 years to reach the billion-person population plateau. Then things sped up considerably. We reached 2 billion people in 1930, about 1000 times faster than it took to reach the first billion. But the rate of increase kept accelerating. By 1950, only 20 years later, we had reached 2.5 billion souls. In 1999, we hit 6 billion. There will be approximately 7 billion people by 2020 and perhaps 11 billion by 2050 to 2100.

Rainforest conversion, which changes forest to fields, and then (usually) to overgrazed, eroded, and infertile land within a generation, is perhaps the most direct executioner of biodiversity. It appears that 25% of the world's topsoil has been lost since 1945. One-third of the world's forest area has disappeared in the same interval. The result is species extinction. A thousand years from now, when humanity reflects on the world that was, and looks out at the desert surrounding the rare and notably less diverse animals that remain, whom will it hold responsible?

President Theodore Roosevelt closed off the Yellowstone region to development in forming the first national park in the United States. Wouldn't it be ironic if some alien equivalent had done the same thing for our planet? Astrobiologists have suggested this—it is known as the Zoo Hypothesis. The joke would be on us: We are somebody's national park, our rare planet Earth stocked with animals for safekeeping. Perhaps that is why we have yet to hear any signals from space. A big fence surrounds our solar system: "Earth Intergalactic Park. Posted: No trespassing or tampering. The only planet with animals for the next 5000 light-years."

****

Twenty thousand years ago, Earth was locked in the glacial grip of the last ice age. Wooly mammoths and great mastodons, ground sloths, camels, and saber-toothed tigers roamed North America; people didn't. Humans were still thousands of years from crossing the land bridge from what would someday be called Siberia to the place we now know as Alaska. Humans were still 10,000 years from mastering agriculture. On some given yet forever anonymous day in that long ago time of 20,000 B.C., a distant neutron star in the constellation Aquila, part of the Summer Triangle so familiar to star watchers in the Northern hemisphere, underwent a violent cataclysm of some type and belched hard radiation into space, hurling an expanding sphere of poison at the speed of light in all directions. For 20,000 years it sped through space. It hit Earth over the Pacific Ocean on the evening of August 27, 1998, as it continued ever onward, its energy diminishing with each mile it traveled from its original source.

For 5 minutes on that late summer day, Earth was bombarded by gamma rays and X-rays, the lethal twins generated by thermonuclear bombs as well as by the interiors of stars. Even after traveling 20,000 light-years, the energy was sufficient to send radiation sensors on seven Earth satellites to maximum reading or off scale. Two of these satellites were shut down to save their instruments from burnout. The radiation penetrated to within 30 miles of Earth's surface and then was dissipated by the lower regions of our planet's atmosphere. This event was the first time that such high energy from outside the solar system was detected to have had a measurable effect on the atmosphere. But in all probability, it was not the first time Earth has been buffeted by energy from interstellar space. Perhaps a closer neutron star, or some other stellar demon not yet known to us, caused one or more of the mass extinctions in Earth's past. Perhaps we have only begun to see the demons surrounding us as we take our first tentative peeks through our planetary bedroom window into space.

Astronomers believe that the 1998 event was caused by the surface disruption of a kind of star that had only been theorized to exist: a magnetar. A type of neutron star, a magnetar is perhaps 20 miles in diameter but is more massive than our sun. It is estimated that a thimbleful of its material would weigh 100 million tons. It is matter compressed far beyond the point of human comprehension. The star has a surface of iron, but iron of a type never found in our solar system. The star spins, as all neutron stars spin, and the result is the formation of an intensely powerful magnetic field. For reasons we can only guess at, the surface of this star—20,000 years ago—underwent a massive disruption, sending energy into space as a consequence.

Energy dissipates with distance. Had the magnetar in question been only 10,000 light-years away, the energy reaching Earth would have been four times stronger—perhaps strong enough to damage the ozone layer. Did this particular event sterilize worlds within a light-year or less? Were there civilizations existing on worlds that were seared out of existence by gamma rays and a magnetic field pulse sufficient to tear the very molecules of living matter apart? Was another Earth sterilized? Perhaps life can flourish only in neighborhoods far from magnetars. Have magnetars—as well as so much else we have seen in the pages of this book—made animal life rare in the Universe? And what else is out there, lurking in the dark?

The discovery of a phenomenon such as the magnetar is an object lesson that suggests a great deal more than life's rarity: There is still so much more to learn about the heavens surrounding us. We humans are like 2-year-olds, just beginning to comprehend the immensity, wonder, and hazards of the wide world. So too with our understanding of astrobiology. It is clearly just beginning.

References

Introduction

Kirschvink, J. L.; Maine, A. T.; and Vali, H. 1997. Paleomagnetic evidence supports a low-temperature origin of carbonate in the Martian meteorite ALH84001. *Science* 275:1629–1633.

Chapter 1. Why Life Might Be Widespread in the Universe

Achenbach-Richter, L.; Gupta, R.; Stetter, K. O.; and Woese, C. R. 1987. Were the original Eubacteria thermophiles? *Systematic and Applied Microbiology* 9:34–39.

Baross, J. A., and Hoffman, S. E. 1985. Submarine hydrothermal vents and associated gradient environments as sites for the origin and evolution of life. *Origins of Life* 15:327–345.

Baross, J. A., and J. W. Deming. 1995. Growth at high temperatures: Isolation, taxonomy, physiology and ecology. In *The microbiology of deep-sea hydrothermal vent habitats*, ed. D. M. Karl, pp. 169–217. Boca Raton, FL: CRC Press.

Baross, J. A., and Holden, J. F. 1996. Overview of hyperthermophiles and their heat-shock proteins. *Advances in Protein Chemistry* 48:1–35.

Caldeira, K., and Kasting, J. F. 1992. Susceptibility of the early Earth to irreversible glaciation caused by carbon ice clouds. *Nature* 359:226–228.

Cech, T. R., and Bass, B. L. 1986. Biological catalysis by RNA. *Annual Review of Biochemistry* 55:599–629.

Chang, S. 1994. The planetary setting of prebiotic evolution. In *Early life on Earth*, Nobel Symposium No. 84, ed. by S. Bengston, pp. 10–23. New York: Columbia Univ. Press.

Doolittle, W. F., and Brown, J. R. 1994. Tempo, mode, the progenote, and the universal root. *Proceedings of the National Academy of Sciences USA* 91:6721–6728.

Doolittle, W. F.; Feng, D. -F.; Tsang, S.; Cho, G.; and Little, E. 1996. Determining divergence times of the major kingdoms of living organisms with a protein clock. *Science* 271:470–477.

Dott, R. H., Jr., and Prothero, D. R. 1993. *Evolution of the Earth.* 5th ed. New York: McGraw-Hill.

Forterre, P. 1997. Protein versus rRNA: Problems in rooting the universal tree of life. *American Society for Microbiology News* 63:89–95.

Forterre, P.; Confalonieri, F.; Charbonnier, F.; and Duguet, M. 1995. Speculations on the origin of life and thermophily: Review of available information on reverse gyrase suggests that hyperthermophilic procaryotes are not so primitive. *Origins of Life and Evolution of the Biosphere* 25:235–249.

Fox, S. W. 1995. Thermal synthesis of amino acids and the origin of life. *Geochimica et Cosmochimica Acta* 59:1213–1214.

Giovannoni, S. J.; Mullins, T. D.; and Field, K. G. 1995. Microbial diversity in oceanic systems: rRNA approaches to the study of unculturable microbes. In *Molecular ecology of aquatic microbes*, ed. I. Joint, pp. 217–248, Berlin: Springer-Verlag.

Glikson, A. Y. 1993. Asteroids and the early Precambrian crustal evolution. *Earth-Science Reviews* 35:285–319.

Gogarten-Boekels, M.; Hilario, E.; and Gogarten, J. P. 1995. *Origins of Life and Evolution of the Biosphere* 25:251–264.

Gold, T. 1998. *The deep hot biosphere.* New York: Springer-Verlag/Copernicus.

Grayling, R. A.; Sandman, K.; and Reeve, J. N. 1996. DNA stability and DNA binding proteins. *Advances in Protein Chemistry* 48:437–467.

Gu, X. 1997. The age of the common ancestor of eukaryotes and prokaryotes: Statistical inferences. *Molecular Biology and Evolution* 14:861–866.

Gupta, R. S., and Golding, G. B. 1996. The origin of the eukaryotic cell. *Trends in Biochemical Sciences* 21:166–171.

Hayes, J. M. 1994. Global methanotrophy at the Archean–Proterozoic transition. In *Early life on Earth.* Nobel Symposium No. 84, ed. S. Bengston, pp. 220–236. New York: Columbia Univ. Press.

Hedén, C.-G. 1964. Effects of hydrostatic pressure on microbial systems. *Bacteriological Reviews* 28:14–29.

Hei, D. J., and Clark, D. S. 1994. Pressure stabilization of proteins from extreme thermophiles. *Applied and Environmental Microbiology* 60:932–939.

Hennet, R.; J.,-C., Holm, N. G.; and Engel, M. H., 1992. Abiotic synthesis of amino acids under hydrothermal conditions and the origin of life: A perpetual phenomenon? *Naturwissenschaften* 79:361–365.

Hilario, E., and Gogarten, J. P. 1993. Horizontal transfer of ATPase genes—the tree of life becomes the net of life. *BioSystems* 31:111–119.

Holden, J. F., and Baross, J. A. 1995. Enhanced thermotolerance by hydrostatic pressure in deep-sea marine hyperthermophile *Pyrococcus* strain ES4. *FEMS Microbiology Ecology* 18:27–34.

Holden, J. F.; Summit, M.; and Baross, J. A. 1997. Thermophilic and hyperthermophilic microorganisms in 3–30°C hydrothermal fluids following a deep-sea volcanic eruption. *FEMS Microbiology Ecology* (in press).

Huber, R.; Stoffers, P.; Hohenhaus, S.; Rachel, R.; Burggraf, S.; Jannasch, H. W.; and Stetter, K. O. 1990. Hyperthermophilic archaeabacteria within the crater and open-sea plume of erupting MacDonald Seamount. *Nature* 345:179–182.

Hunten, D. M. 1993. Atmospheric evolution of the terrestrial planets. *Science* 259:915–920.

Kadko, D.; Baross, J.; and Alt, J. 1995. The magnitude and global implications of hydrothermal flux. In *Physical, chemical, biological and geological interactions within sea floor hydrothermal discharge*, Geophysical Monograph 91, ed. S. Humphris, R. Zierenberg, L. Mullineaux, and R. Thompson, pp. 446–466. Washington, DC: AGU Press.

Karhu, J., and Epstein, S. 1986. The implication of the oxygen isotope records in coexisting cherts and phosphates. *Geochimica et Cosmochimica Acta* 50:1745–1756.

Kasting, J. F. 1984. Effects of high CO_2 levels on surface temperature and atmospheric oxidation state of the early Earth. *Journal of Geophysical Research* 86:1147–1158.

Kasting, J. F. 1993. New spin on ancient climate. *Nature* 364:759–760.

Kasting, J. F. 1997. Warming early Earth and Mars. *Science* 276:1213–1215.

Kasting, J. F., and Ackerman, T. P. 1986. Climatic consequences of very high carbon dioxide levels in the Earth's early atmosphere. *Science* 234:1383–1385.

Knauth, L. P., and Epstein, S. 1976. Hydrogen and oxygen isotope ratios in nodular and bedded cherts. *Geochimica et Cosmochimica Acta* 40:1095–1108.

Knoll, A. 1998. A Martian chronicle. *The Sciences* 38:20–26.

Lazcano, A. 1994. The RNA world, its predecessors, and its descendants. In *Early life on Earth*, Nobel Symposium No. 84, ed. S. Bengston, pp. 70–80. New York: Columbia Univ. Press.

L'Haridon, S. L.; Raysenbach, A.-L.; Glénat, P.; Prieur, D.; and Jeanthon, C. 1995. Hot subterranean biosphere in a continental oil reservoir. *Nature* 377:223–224.

Lowe, D. R. 1994. Early environments: Constraints and opportunities for early evolution. In *Early life on Earth*, Nobel Symposium No. 84, ed. S. Bengston, pp. 24–35. New York: Columbia Univ. Press.

Maher, K. A., and Stevenson, J. D. 1988. Impact frustration of the origin of life. *Nature* 331:612–614.

Marshall, W. L. 1994. Hydrothermal synthesis of amino acids. *Geochimica et Cosmochimica Acta* 58:2099–2106.

Michels, P. C., and Clark, D. S. 1992. Pressure dependence of enzyme catalysis. In *Biocatalysis at extreme environments*, ed. M. W. W. Adams and R. Kelly, pp. 108–121. Washington, DC: American Chemical Society Books.

Miller, S. L. 1953. A production of amino acids under possible primitive Earth conditions. *Science* 117:528–529.

Miller, S. L., and Bada, J. L. 1988. Submarine hot springs and the origin of life. *Nature* 334:609–611.

Mojzsis, S.; Arrhenius, G.; McKeegan, K. D.; Harrison, T. M.; Nutman, A. P.; and Friend, C. R. L. 1966. Evidence for life on Earth before 3,800 million years ago. *Nature* 385:55–59.

Moorbath, S.; O'Nions, R. K.; and Pankhurst, R. J. 1973. Early Archaean age of the Isua iron formation. *Nature* 245:138–139.

Newman, M. J., and Rood, R. T. 1977. Implications of solar evolution for the Earth's early atmosphere. *Science* 198:1035–1037.

Nickerson, K. W. 1984. An hypothesis on the role of pressure in the origin of life. *Theoretical Biology* 110:487–499.

Nisbet, E. G. 1987. The young Earth: An introduction to Archaean geology. Boston: Allen & Unwin.

Nutman, A. P.; Mojzsis, S. J.; and Friend, C. R. L. 1997. Recognition of ≥ 3850 Ma water-lain sediments in West Greenland and their significance for the early Archaean Earth. *Geochimica et Cosmochimica Acta* 61:2475–2484.

Oberbeck, V. R., and Mancinelli, R. L. 1994. Asteroid impacts, microbes, and the cooling of the atmosphere. *BioScience* 44:173–177.

Oberbeck, V. R.; Marshall, J. R.; and Aggarwal, H. R. 1993. Impacts, tillites, and the breakup of Gondwanaland. *Journal of Geology* 101:1–19.

Ohmoto, H., and Felder, R. P. 1987. Bacterial activity in the warmer, sulphate-bearing Archaean oceans. *Nature* 328:244–246.

Pace, N. 1991. Origin of life—facing up to the physical setting. *Cell* 65:531–533.

Perry, E. C., Jr.; Ahmad, S. N.; and Swulius, T. M. 1978. The oxygen isotope composition of 3,800 m.y. old metamorphosed chert and iron formation from Isukasia West Greenland. *Journal of Geology* 86:223–239.

Sagan, C., and Chyba, C. 1997. The early faint sun paradox: Organic shielding of ultraviolet-labile greenhouse gases. *Science* 276:1217–1221.

Schidlowski, M. 1988. A 3,800-million-year isotopic record of life from carbon in sedimentary rocks. *Nature* 333:313–318.

Schidlowski, M. 1993. The initiation of biological processes on Earth: Summary of empirical evidence. In *Organic Geochemistry*, ed. M. H. Engel and S. A. Macko, pp. 639–655. New York: Plenum Press.

Schopf, J. W. 1994. The oldest known records of life: Early Archean stromatolites, microfossils, and organic matter. In *Early life on Earth*. Nobel Symposium No. 84, ed. S. Bengston, pp. 193–206. New York: Columbia Univ. Press.

Schopf, J. W., and Packer, B. M. 1987. Early Archean (3.3-billion to 3.5-billion-year-old) microorganisms from the Warrawoona Group, Australia. *Science* 237:70–73.

Shock, E. L. 1992. Chemical environments of submarine hydrothermal systems. *Origin of Life and Evolution of the Biosphere* 22:67–107.

Sogin, M. L. 1991. Early evolution and the origin of eukaryotes. *Current Opinion in Genetics and Development* 1:457–463.

Sogin, M. L.; Silverman, J. D.; Hinkle, G.; and Morrison, H. G. 1996. Problems with molecular diversity in the Eucarya. In *Society for General Microbiology Symposium: Evolution of microbial life*, ed. D. M. Roberts, P. Sharp, G. Alderson, and M. A. Collins, pp. 167–184. Cambridge, England: Cambridge Univ. Press.

Staley, J. T., and J. J. Gosink. Poles apart: Biodiversity and biogeography of polar sea ice bacteria. *Ann. Rev. Microbiol.* (in press).

Stevens, T. O., and McKinley, J. P. 1995. Lithoautotrophic microbial ecosystems in deep basalt aquifers. *Science* 270:450–454.

Woese, C. R. 1994. There must be a prokaryote somewhere: Microbiology's search for itself. *Microbiological Reviews* 58:1–9.

Woese, C. R.; Kandler, O.; and Wheelis, M. L. 1990. Towards a natural system of organisms: Proposals for the domains Archaea, Bacteria, and Eucarya. *Proceedings of the National Academy of Sciences USA* 87:4576–4579.

Chapter 2. Habitable Zones of the Universe

Cloud, P. 1987. *Oasis in space.* New York: Norton.

De Duve, C. 1995. *Vital Dust.* New York: Basic Books.

Dole, S. 1964. *Habitable planets for man.* New York: Blaisdell.

Doolittle, W. F. 1999. Phylogenetic classification and the Universal Tree. *Science* 284:2124–2128.

Doyle, L. R. 1996. Circumstellar habitable zones, *Proceedings of the First International Conference.* Menlo Park, CA: Travis House.

Forget, F., and Pierrehumbert, G. D. 1997. Warming early Mars with carbon dioxide that scatters infrared radiation. *Science* 278:1273–1276.

Hart, M. 1978. The evolution of the atmosphere of the earth. *Icarus* 33:23–39.

Hart, M. H. 1979. Habitable zones about main sequence stars. *Icarus* 37:351–357.

Illes-Almar, E.; Almar, I.; Berczi, S.; and Likacs, B. 1997. On a broader concept of circumstellar habitable zones. Conference Paper, Astronomical and Biochemical Origins and the Search for Life in the Universe, IAU Colloquium 161, Bologna, Italy, p. 747.

Kasting, J. F. 1988. Runaway and moist greenhouse atmospheres and the evolution of Earth and Venus. *Icarus* 74:472–494.

Kasting, J. F. 1993. Earth's early atmosphere. Science 259:920–926.

Kasting, J. F. 1997. Habitable zones around low mass stars and the search for extraterrestrial life. In *Planetary and interstellar processes relevant to the origins of life,* ed. D. C. B. Whittet, p. 291. Kluwer Academic Publishers, 1997.

Kasting, J. F. 1997. Update: The early Mars climate question heats up. *Science* 278:1245.

Kasting, J. F.; Whitmire, D. P.; and Reynolds, R. T. 1993. Habitable zones around main sequence stars. *Icarus* 101:108–128.

Ksanfomaliti, L. V. 1998. Planetary systems around stars of late spectral types: A Limitation for habitable zones. *Astronomicheskii Vestnik* 32:413.

Lepage, A. J. 1998. Habitable moons. *Sky and Telescope* 96: 50.

Miller, S. L. 1953. Production of amino acids under possible primitive Earth conditions. *Science* 117:528.

Sagan, C., and Chyba, C. 1997. The early faint sun paradox: Organic shielding of ultraviolet-labile greenhouse gases. *Science* 276:1217–1221.

Sleep, N. H.; Zahnle, K. J.; Kasting, J. F.; and Morowitz, H. J. 1989. Annihilation of ecosystems by large asteroid impacts on the early Earth. *Nature* 342:139.

Squyres, S. W., and Kasting, J. F., 1994. Early Mars—how warm and how wet? *Science* 265, 744.

Wetherill, G. W. 1996. The formation and habitability of extra-solar planets. *Icarus* 119:219–238.

Whitmire, D. P., Matese, J. J.; Criswell, L.; and Mikkola, S. 1998. Habitable planet formation in binary star systems. *Icarus* 132:196–203.

Williams, D. M., Kasting, J. F.; and Wade, R. A. 1996. Habitable moons around extrasolar giant planets. AAS/Division of Planetary Sciences Meeting 28, 1221.

Williams, D. M., Kasting, J. F.; and Wade, R. A. 1997. Habitable moons around extrasolar giant planets. *Nature* 385: 234–236.

Chapter 3. Building a Habitable Earth

Bryden, G., D.; Lin, N. C.; and Terquem, C. 1998. Planet formation; orbital evolution and planet-star tidal interaction. ASP Conf. Ser. 138: 1997 Pacific Rim Conference on Stellar Astrophysics 23.

Cameron, A. G. W. 1995. The first ten million years in the solar nebula. *Meteoritics* 30, 133–161.

Chyba, C. F. 1987. The cometary contribution to the oceans of the primitive Earth. *Nature* 220:632–635.

Chyba, C. F. 1993. The violent environment of the origin of life: Progress and uncertainties. *Geochimica et Cosmochimica Acta* 57:3351–3358.

Chyba, C. F., and Sagan, C. 1992. Endogenous production, exogenous delivery, and impact-shock synthesis of organic molecules: An inventory for the origins of life. *Nature* 355:125–131.

Chyba, C. F.; Thomas, P. J.; Brookshaw, L.; and Sagan, C. 1990. Cometary delivery of organic molecules to the early Earth. *Science* 249:366–373.

Holland, H. D. 1984. *The chemical evolution of the atmosphere and oceans.* Princeton, NJ: Princeton Univ. Press.

Lin, D. N. C. 1997. On the ubiquity of planets and diversity of planetary systems. Proceedings of the 21st Century Chinese Astronomy Conference: dedicated to Prof. C. C. Lin, Hong Kong, 1–4 August 1996, ed. K. S. Cheng and K. L. Chan, Singapore. River Edge, NJ: World, Scientific, p. 313.

Lunine, J., 1999. *Earth: Evolution of a habitable world.* Cambridge, England: Cambridge Univ. Press.

Maher, K. A. J., and Stevenson, D. J. 1988. Impact frustration of the origin of life. *Nature* 331:612–614.

Sagan, C., and Chyba, C. 1997. The early faint sun paradox: Organic shielding of ultraviolet-labile greenhouse gases. *Science* 276:1217–1221.

Sleep, N. H.; Zahnle, K. J.; Kasting, J. F.; and Morowitz, H. J. 1989. Annihilation of ecosystems by large asteroid impacts on the early Earth. *Nature* 342:139–142.

Taylor, S. R. 1998. On the difficulties of making earth-like planets. *Meteoritics and Planetary Science* 32:153.

Taylor, S. R., and McLennan, S. M. 1995. The geochemical evolution of the continental crust. *Reviews in Geophysics* 33:241–265.

Towe, K. M. 1994. Earth's early atmosphere: Constraints and opportunities for early evolution. In *Early life on Earth*, Nobel Symposium No. 84, ed. S. Bengston, pp. 36–47. New York: Columbia Univ. Press.

van Andel, T. H. 1985. *New views on an old planet.* Cambridge, England: Cambridge Univ. Press.

Walker, J. C. G. 1977. *Evolution of the atmosphere.* London: Macmillan.

Wetherill, G. W. 1991. Occurrence of Earth-like bodies in planetary systems. *Science* 253:535–538.

Wetherill, G. W. 1994. Provenance of the terrestrial planets. *Geochimica et Cosmochimica Acta* 58:4513–4520.

Wetherill, G. W. 1996. The formation and habitability of extra-solar planets. *Icarus* 119:219–238.

Chapter 4. Life's First Appearance on Earth

Abbott, D. H., and Hoffman, S. E. 1984. Archaean plate tectonics revisited. 1. Heat flow, spreading rate, and the age of subducting oceanic lithosphere and their effects on the origin and evolution of continents. *Tectonics* 3:429–448.

Bada, J. L.; Bigham, C.; and Miller, S. L. 1994. Impact melting of frozen oceans on the early Earth: Implications for the origin of life. *Proceedings of the National Academy of Sciences USA* 91:1248–1250.

Barns, S. M.; Fundyga, R. E.; Jeffries, M. W.; and Pace, N. R. 1994. Remarkable archaeal diversity detected in a Yellowstone National Park hot spring environment. *Proceedings of the National Academy of Sciences USA* 91:1609–1613.

Baross, J. A., and Deming, J. W. 1995. Growth at high temperatures: Isolation and taxonomy, physiology, and ecology. In *The microbiology of deep-sea hydrothermal vent habitats*, ed. D. M. Karl, pp. 169–217. Boca Raton, FL: CRC Press.

Baross, J. A., and Hoffman, S. E. 1985. Submarine hydrothermal vents and associated gradient environments as sites for the origin and evolution of life. *Orig. Life Evolution Biosphere* 15:327–345.

Brakenridge, G. R.; Newsom, H. E.; and Baker, V. R. 1985. Ancient hot springs on Mars: Origins and paleoenvironmental significance of small Martian valleys. *Geology* 13:859–862.

Carl, M. H. 1996. *Water on Mars*. New York: Oxford Univ. Press.

Converse, D. R.; Holland, H. D.; and Edmond, J. M. 1984. Flow rates in the axial hot springs of the East Pacific Rise (21°N): Implications for the heat budget and the formation of massive sulfide deposits. *Earth Planet. Sci. Lett.* 69:159–175.

Criss, R. E., and Taylor, H. P., Jr. 1986. Meteoric-hydrothermal systems. *Rev. Mineral.* 16:373–424.

Daniel, R. M. 1992. Modern life at high temperatures. In Marine Hydrothermal Systems and the Origin of Life, ed. N. Holm, *Orig. Life Evolution Biosphere* 22:33–42.

Doolittle, W. F. 1999. Phylogenetic classification and the Universal Tree. *Science* 284:2124.

Glikson, A. 1995. Asteroid comet mega-impacts may have triggered major episodes of crustal evolution. *Eos*, 76:49–54.

Griffith, L. L., and Shock, E. L. 1995. A geochemical model for the formation of hydrothermal carbonate on Mars. *Nature* 377:406–408.

Griffith, L. L., and Shock, E. L. 1997. Hydrothermal hydration of Martian crust: Illustration via geochemical model calculations. *J. Geophys. Res.* 102:9135–9143.

Karl, D. M. 1995. Ecology of free-living, hydrothermal vent microbial communities. In *The microbiology of deep-sea hydrothermal vent habitats*, ed. D. M. Karl, pp. 35–124. Boca Raton, FL: CRC Press.

MacLeod, G.; McKeown, C.; Hall, A. J.; and Russell, M. J. 1994. Hydrothermal and oceanic pH conditions of possible relevance to the origin of life. *Orig. Life Evolution Biosphere* 23:19–41.

McCollom, T. M., and Shock, E. L. 1997. Geochemical constraints on chemolithoautotrophic metabolism by microorganisms in seafloor hydrothermal systems. *Geochimica et Cosmochimica Acta* (in press).

McSween, Jr., H. Y. 1994. What we have learned about Mars from SNC meteorites. *Meteoritics* 29:757–779.

Miller, S., and Lazcano, A. 1996. From the primitive soup to Cyanobacteria: It may have taken less than 10 million years. In *Circumstellar habitable zones*, ed. L. Doyle, pp. 393–404. Menlo Park, CA: Travis House.

Pace, N. R. 1991. Origin of life—facing up to the physical setting. *Cell* 65:531–533.

Romanek, C. S.; Grady, M. M.; Wright, I. P.; Mittlefehldt, D. W.; Socki, R. A.; C. T. Pillinger, C. T.; and Gibson, Jr., E. K. 1994. Record of fluid rock interactions on Mars from the meteorite ALH84001. *Nature* 372:655–657.

Russell, M. J.; Daniel, R. M.; and Hall, A. J. 1993. On the emergence of life via catalytic iron sulphide membranes. *Terra Nova* 5:343–347.

Russell, M. J.; Daniel, R. M.; Hall, A. J.; and Sherringham, J. 1994. A hydrothermally precipitated catalytic iron sulphide membrane as a first step toward life. *J. Molec. Evol.* 39:231–243.

Russell, M. J., and Hall, A. J. 1995. The emergence of life at hot springs: A basis for understanding the relationships between organics and mineral deposits. In *Proceedings of the Third Biennial SGA Meeting, Prague, Mineral deposits: From their origin to their environmental impacts,* ed. J. Pasava, B. Kribek, and K. Zak, pp. 793–795.

Russell, M. J., and Hall, A. J. 1997. The emergence of life from iron mono-sulphide bubbles at a hydrothermal redox front. *J. Geol. Soc.* (in press).

Russell, M. J.; Hall, A. J.; Cairns-Smith, A. G.; and Braterman, P. S. 1988. Submarine hot springs and the origin of life. *Nature* 336:117.

Russell, M. J.; Hall, A. J.; and Turner, D. 1989. *In vitro* growth of iron sulphide chimneys: Possible culture chambers for origin-of-life experiments. *Terra Nova* 1:238–241.

Schwartzman, D.; McMenamin, M.; and Volk, T. 1993. Did surface temperatures constrain microbial evolution? *BioScience* 43:390–393.

Seewald, J. S. 1994. Evidence for metastable equilibrium between hydrocarbons under hydrothermal conditions. *Nature* 370:285–287.

Segerer, A. H.; Burggraf, S.; Fiala, G.; Huber, G.; Huber, R.; Pley, U.; and Stetter, K. O. 1993. Life in hot springs and hydrothermal vents. *Orig. Life Evol. Biosphere* 23:77–90.

Shock, E. L. 1990a. Geochemical constraints on the origin of organic compounds in hydrothermal systems. *Orig. Life Evol. Biosphere* 20:331–367.

Shock, E. L. Chemical environments in submarine hydrothermal systems. 1992a. In Holm, N. Marine hydrothermal systems and the origin of life, ed. N. Holm. *Orig. Life Evol. Biosphere* 22:67–107.

Shock, E. L.; McCollom, T.; and Schulte, M. D. 1995. Geochemical constraints on chemolithoautotrophic reactions in hydrothermal systems. *Orig. Life Evol. Biosphere* 25:141–159.

Shock, E. L., and Schulte, M. D. 1997. Hydrothermal systems as locations of organic synthesis on the early Earth and Mars. *Orig. Life Evol. Biosphere* (in press).

Sleep, N. H.; Zahnle, K. J.; Kasting, J. F.; and Morowitz, H. J. 1989. Annihilation of ecosystems by large asteroid impacts on the early Earth. *Nature* 342:139–142.

Stetter, K. O. 1995. Microbial life in hyperthermal environments. *ASM News, American Society for Microbiology* 61:285–290.

Treiman, A. H. 1995. A petrographic history of Martian meteorite ALH84001: Two shocks and an ancient age. *Meteoritics* 30:294–302.

Von Damm, K. L. 1990. Seafloor hydrothermal activity: Black smoker chemistry and chimneys. *Ann. Rev. Earth Planet. Sci.* 18:173–204.

Watson, L. L.; Hutcheon, I. D.; Epstein, S.; and Stolper, E. M. 1994. Water on Mars: Clues from deuterium/hydrogen and water contents of hydrous phases in SNC meteorites. *Science* 265:86–90.

Wilson, E. 1992. *The diversity of life.* Cambridge, MA: Harvard Univ. Press.

Wilson, L., and Head, III, J. W. 1994. Mars: Review and analysis of volcanic eruption theory and relationships to observed landforms. *Rev. Geophys.* 32:221–263.

Woese, C. R. 1987. Bacterial evolution. *Microbiol. Rev.* 51:221–271.

Woese, C. R.; Kandler, O.; and Wheelis, M. L. 1990. Towards a natural system of organisms: Proposal for the domains Archaea, Bacteria, and Eucarya. *Proceedings of the National Academy of Sciences USA* 87:4576–4579.

Chapter 5. How to Build Animals

Akam, M., *et al.*, eds. 1994. *The evolution of developmental mechanisms.* Cambridge, England: The Company of Biologists, Ltd.

Brasier, M. D.; Shields, G.; Kuleshoy, V. N.; and Zhegallos, E. A. 1996. Integrated chemo- and biostratigraphic calibration of early animal evolution: Neoproterozoic-early Cambrian of southwest Mongolia. *Geological Magazine* 133:445–485.

Bowring, S. A., Grotzinger, J. P.; Isachsen, C. E.; Knoll, A. H.; Pelechaty, S. M.; and Kolosov, P. 1993. Calibrating rates of Early Cambrian evolution. *Science* 261:1293–1298.

Carroll, S. B. 1995. Homeotic genes and the evolution of arthropods and chordates. *Nature* 376:479–485.

Chen, J.-Y., and Erdtmann, B.-D. 1991. Lower Cambrian fossil lagerstatte from Chengjiang, Yunnan, China: Insights for reconstructing early metazoan life. In *The early evolution of metazoa and the significance of problematic taxa*, ed. A. M. Simonetta and S. Conway Morris, pp. 57–76. Cambridge, England: Cambridge Univ. Press.

Conway Morris, S. 1997. Defusing the Cambrian "explosion"? *Current Biology* 7:R71–R74.

Crimes, T. P. 1994. The period of early evolutionary failure and the dawn of evolutionary success: The record of biotic changes across the Precambrian–Cambrian boundary. In *The paleobiology of trace fossils*, ed. S. K. Donovan, pp. 105–133. London: Wiley.

Erwin, D. H. 1993. The origin of metazoan development. *Biological Journal of the Linnean Society* 50:255–274.

Evans, D. A. 1998. True polar wander, a supercontinental legacy. *Earth and Planetary Science Letters* 157:1–8.

Evans, D. A.; Beukes, N. J.; and Kirschvink, J. L. 1997. Low-latitude glaciation in the Paleoproterozoic era. *Nature* 386(6622):262–266.

Evans, D. A.; Ripperdan, R. L.; and Kirschvink, J. L. 1998. Polar wander and the Cambrian (response). *Science* 279:16.
Full article accessible at http://www.sciencemag.org/cgi/content/full/279/5347/9a

Evans, D. A.; Zhuravlev, A. Y.; Budney, C. J.; and Kirschvink, J. L. 1996. Paleomagnetism of the Bayan Gol Formation, western Mongolia. *Geological Magazine* 133:478–496.

Fedonkin, M. A., and B. M. Waggoner. 1996. The Vendian fossil *Kimberella*: The oldest mollusk known. *Geological Society of America, Abstracts with Program*. 28(7):A–53.

Grotzinger, J. P.; Bowring, S. A.; Saylor, B.; and Kauffman, A. J. 1995. New biostratigraphic and geochronological constraints on early animal evolution. *Science* 270:598–604.

Kappen, C.; and Ruddle, F. H. 1993. Evolution of a regulatory gene family: *HOM/Hox* genes. *Current Opinion in Genetics and Development* 3:931–938.

Knoll, A., and Carroll, S. 1999. Early animal evolution: Emerging views from comparative biology and geology. *Science* 284:2129–2137.

Knoll, A. H.; Kaufman, A. J.; Semikhatov, M. A.; Grotzinger, J. P.; and Adams, W. 1995. Sizing up the sub-Tommotian unconformity in Siberia. *Geology* 23:1139–1143.

Margulis, L., and Sagan, D. 1986. *Microcosmos*. New York: Simon & Schuster.

Raff, R. A. 1996. *The shape of life*. Chicago: Univ. of Chicago Press.

Schwartzman, D., and Shore, S. 1996. Biotically mediated surface cooling and habitability for complex life. In *Circumstellar habitable zones*, ed. L. Doyle, pp. 421–443. Menlo Park, CA: Travis House.

Valentine, J. W. 1994. Late Precambrian bilaterans: Grades and clades. *Proceedings of the National Academy of Sciences* 91:6751–6757.

Valentine, J. W.; Erwin, D. H.; and Jablonski, D. 1996. Developmental evolution of metazoan body plans: The fossil evidence. *Developmental Biology* 173:373–381.

Wilmer, P. 1990. *Invertebrate relationships: Patterns in animal evolution*. Cambridge, England: Cambridge Univ. Press.

Chapter 6. Snowball Earth

Bertani, L. E.; Huang, J.; Weir, B.; and Kirschvink, J. L. 1997. Evidence for two types of subunits in the bacterioferritin of *Magnetospirillum magnetotacticum*. *Gene* 201:31–36.

Evans, D. A.; Beukes, N. J.; and Kirschvink, J. L. 1997. Low-latitude glaciation in the Paleoproterozoic era. *Nature* 386(6622):262–266.

Evans, D. A.; Zhuravlev, A. Y.; Budney, C. J.; and Kirschvink, J. L. 1996. Paleomagnetism of the Bayan Gol Formation, western Mongolia. *Geological Magazine* 133:478–496.

Hoffman, P.; Kaufman, A.; Halverson, G.; and Schrag, D. 1998. A Neoproterozoic Snowball Earth. *Science* 281:1342–1346.

Kirschvink, J. L. 1992. A paleogeographic model for Vendian and Cambrian time. In *The Proterozoic biosphere: A multidisciplinary study*, ed. J. W. Schopf, C. Klein, and D. Des Maris, pp. 567–581. Cambridge, England: Cambridge Univ. Press.

Kirschvink, J. L.; Gaidos, E. J.; Bertani, L. E.; Beukes, N. J.; Gutzmer, J.; Evans, D. A.; Maepa, L. N.; and Steinberger, R. E. The paleoproterozoic snowball Earth: deposition of the Kalahari manganese field and evolution of the Archaea and Eukarya kingdoms. *Science*, in extended review (as of 11/98).

Schwartzman, D.; McMenamin, M.; and Volk, T. 1993. Did surface temperatures constrain microbial evolution? *BioScience* 43:390–393.

Chapter 7. The Enigma of the Cambrian Explosion

Aitken, J. D., and McIlreath, I. A. 1984. The Cathedral Reef Escarpment, a Cambrian great wall with humble origins. *Geos* 13:17–19.

Allison, P. A., and Brett, C. E. 1995. *In situ* benthos and paleo-oxygenation in the Middle Cambrian Burgess Shale, British Columbia, Canada. *Geology* 23:1079–1082.

Aronson, R. B. 1992. Decline of the Burgess Shale fauna: Ecologic or taphonomic restriction? *Lethaia* 25:225–229.

Bergström, J. 1986. *Opabinia* and *Anomalocaris*, unique Cambrian "arthropods." *Lethaia* 19:241–246.

Briggs, D. E. G. 1979. *Anomalocaris*, the largest known Cambrian arthropod. *Palaeontology* 22:631–664.

Briggs, D. E. G. 1992. Phylogenetic significance of the Burgess Shale crustacean *Canadaspis*. *Acta Zoologica (Stockholm)* 73:293–300.

Briggs, D. E. G., and Collins, D. 1988. A Middle Cambrian chelicerate from Mount Stephen, British Columbia. *Palaeontology* 31:779–798.

Briggs, D. E. G., and Fortey, R. A. 1989. The early radiation and relationships of the major arthropod groups. *Science* 246:241–243.

Briggs, D. E. G., and Whittington, H. B. 1985. Modes of life of arthropods from the Burgess Shale, British Columbia. *Philosophical Transactions of the Royal Society of Edinburgh* 76:149–160.

Budd, G. E. 1996. The morphology of *Opabinia regalis* and the reconstruction of the arthropod stem-group. *Lethaia* 29:1–14.

Butterfield, N. J. 1990a. Organic preservation of non-mineralizing organisms and the taphonomy of the Burgess Shale. *Paleobiology* 16:272–286.

Butterfield, N. J. 1997. Plankton ecology and the Proterozoic–Phanerozoic transition. *Paleobiology* 23:247–262.

Butterfield, N. J., and Nicholas, C. J. 1996. Burgess Shale-type preservation of both non-mineralizing and "shelly" Cambrian organisms from the Mackenzie Mountains, northwestern Canada. *Journal of Paleontology* 70:893–899.

Chen Junyuan; Edgecombe, G. D.; Ramsköld, L.; and Zhou Guiqing. 1995. Head segmentation in early Cambrian *Fuxianhuia*: Implications for arthropod evolution. *Science* 268:1339–1343.

Chen Junyuan; Edgecombe, G. D.; and Ramsköld, L. 1997. Morphological and ecological disparity in naraoiids (Arthropoda) from the Early Cambrian Chengjiang fauna, China. *Records of the Australian Museum* 49:1–24.

Chen Junyuan; Ramsköld, L.; and Zhou Guiqing. 1994. Evidence for monophyly and arthropod affinity of Cambrian predators. *Science* 264:1304–1308.

Chen Junyuan; Zhou Guiqing; Zhu Maoyan; and Yeh K. Y. ca. 1996. *The Chengjiang biota. A unique window on the Cambrian explosion.* National Museum of Natural Science, Taiwan. [in Chinese]

Cloud, P. 1987. *Oasis in space.* New York: Norton.

Collins, D.; Briggs, D.; and Conway Morris, S. 1983. New Burgess Shale fossil sites reveal Middle Cambrian faunal complex. *Science* 222:163–167.

Conway Morris, S. 1979a. The Burgess Shale (Middle Cambrian) fauna. *Annual Review of Ecology and Systematics* 10:327–349.

Conway Morris, S., ed. 1982. *Atlas of the Burgess Shale.* London: Palaeontological Association.

Conway Morris, S. 1989. Burgess Shale faunas and the Cambrian explosion. *Science* 246:339–346.

Conway Morris, S. 1989. The persistence of Burgess Shale-type faunas: Implications for the evolution of deeper-water faunas. *Transactions of the Royal Society of Edinburgh: Earth Sciences* 80:271–283.

Conway Morris, S. 1990. Late Precambrian and Cambrian soft-bodied faunas. *Annual Review of Earth and Planetary Sciences* 18:101–22.

Conway Morris, S. 1992. Burgess Shale-type faunas in the context of the "Cambrian explosion": A review. *Journal of the Geological Society, London* 149:631–636.

Conway Morris, S. 1993a. Ediacaran-like fossils in Cambrian Burgess Shale-type faunas of North America. *Palaeontology* 36:593–635.

Conway Morris, S. 1993b. The fossil record and the early evolution of the metazoa. *Nature* 361:219–225.

Conway Morris, S. 1998. *Crucible of creation.* Oxford Univ. Press.

Conway Morris, S., and Whittington, H. B. 1985. Fossils of the Burgess Shale, a national treasure in Yoho National Park, British Columbia. *Miscellaneous Reports of the Geological Survey of Canada* 43:1–31.

Dzik, J. 1995. *Yunnanozoon* and the ancestry of chordates. *Acta Palaeontologica Polonica* 40:341–360.

Erwin, D. M. 1993. The origin of metazoan development: A palaeobiological perspective. *Biological Journal of the Linnean Society* 50:255–274.

Fritz, W. H. 1971. Geological setting of the Burgess Shale. In *Symposium on Extraordinary Fossils. Proceedings of the North American Paleontological Convention*, Field Museum of Natural History, Chicago. September 5–7, 1969, Part I, pp. 1155–1170. Lawrence, KS: Allen Press.

Gould, S. J. 1986. *Wonderful life.* New York: Norton.

Gould, S. J. 1991. The disparity of the Burgess Shale arthropod fauna and the limits of cladistic analysis: Why we must strive to quantify morphospace. *Paleobiology* 17:411–423.

Grotzinger, J. P.; Bowring, S. A.; Saylor, B. Z.; and Kaufman, A. J. 1995. Biostratigraphic and geochronologic constraints on early animal evolution. *Science* 270:598–604.

Kirschvink, J. L.; Magaritz, M.; Ripperdan, R. L.; Zhuravlev, A. Y.; and Rozanov, A. Y. 1991. The Precambrian–Cambrian boundary: Magnetostratigraphy and carbon isotopes resolve correlation problems between Siberia, Morocco, and South China. *GSA Today* 1:69–91.

Kirschvink, J. L.; Ripperdan, R. L.; and Evans, D. A. 1997. Evidence for a large-scale Early Cambrian reorganization of continental masses by inertial interchange true polar wander. *Science* 277:541–545.

Kirschvink, J. L., and Rozanov, A. Y. 1984. Magnetostratigraphy of Lower Cambrian strata from the Siberian Platform: A paleomagnetic pole and a preliminary polarity time scale. *Geological Magazine* 121:189–203.

Ludvigsen, R. 1989. The Burgess Shale: Not in the shadow of the Cathedral Escarpment. *Geoscience Canada* 16:51–59.

McMenamin, M., and McMenamin, R. 1990. *The emergence of animals.* New York: Columbia Univ. Press.

Ramsköld, L., and Hou Xianguang. 1991. New early Cambrian animal and onychophoran affinities of enigmatic metazoans. *Nature* 351:225–228.

Rigby, J. K. 1986. Sponges of the Burgess Shale (Middle Cambrian), British Columbia. *Palaeontographica Canadiana* 2:1–105.

Simonetta, A. M., and Conway Morris, S., eds. 1991. *The early evolution of metazoa and the significance of problematic taxa.* Cambridge, England: Cambridge Univ. Press.

Simonetta, A. M., and Insom, E. 1993. New animals from the Burgess Shale (Middle Cambrian) and their possible significance for the understanding of the Bilateria. *Bollettino Zoologica* 60:97–107.

Towe, K. M. 1996. Fossil preservation in the Burgess Shale. *Lethaia* 29:107–108.

Whittington, H. B. 1971a. The Burgess Shale: History of research and preservation of fossils. In *Symposium on extraordinary fossils. Proceedings of the North American Paleontological Convention,* Field Museum of Natural History, Chicago, September 5–7, 1969, Part I, pp. 1170–1201. Lawrence, KS: Allen Press.

Whittington, H. B. 1979. Early arthropods, their appendages and relationships. In *The origin of major invertebrate groups,* ed. M. R. House. Systematics Association Special Volume 12, pp. 253–268.

Whittington, H. B., and Briggs, D. E. G. 1985. The largest Cambrian animal, *Anomalocaris,* Burgess Shale, British Columbia. *Philosophical Transactions of the Royal Society of London* B 309:569–609.

Wills, M. A.; Briggs, D. E. G.; and Fortey, R. A. 1994. Disparity as an evolutionary index: A comparison of Cambrian and Recent arthropods. *Paleobiology* 20:93–130.

Yochelson, E. L. 1996. Discovery, collection, and description of the Middle Cambrian Burgess Shale biota by Charles Doolittle Walcott. *Proceedings of the American Philosophical Society* 140:469–545.

Chapter 8. Mass Extinctions and the Rare Earth Hypothesis

Alvarez, L.; Alvarez, W.; Asaro, F.; and Michel, H. 1980. Extra-terrestrial cause for the Cretaceous–Tertiary extinction. *Science* 208:1094–1108.

Bourgeois, J. 1994. Tsunami deposits and the K/T boundary: A sedimentologist's perspective. *Lunar Planetary Institute Cont.* 825:16.

Caldeira, K., and Kasting, J. F. 1992. Susceptibility of the early Earth to irreversible glaciation caused by carbon ice clouds. *Nature* 359:226–228.

Covey, C.; Thompson, S.; Weissman, P.; and MacCracken, M. 1994. Global climatic effects of atmospheric dust from an asteroid or comet impact on earth. *Global and Planetary Change* 9:263–273.

Dar, A.; Laor, A.; and Shaviv, N. 1998. Life extinctions by cosmic ray jets. *Physical Rev. Let.* 80:5813–5816.

Donovan, S. 1989. *Mass extinctions: Processes and evidence.* New York: Columbia Univ. Press.

Ellis, J., and Schramm, D. 1995. Could a supernova explosion have caused a mass extinction? *Proc. Nat. Acad. Sci.* 92:235–238.

Erwin, D. 1993. The great Paleozoic crisis: Life and death in the Permian. New York: Columbia Univ. Press.

Erwin, D. 1994. The Permo-Triassic extinction. *Nature* 367:231–236.

Grieve, R. 1982. The record of impact on Earth. Geol. Soc. America Special Paper 190, ed. Silver, S., and Schultz, P., pp. 25–37.

Hallam, A. 1994. The earliest Triassic as an anoxic event, and its relationship to the End-Paleozoic mass extinction. In *Global environments and resources,* pp. 797–804. Canadian Society of Petroleum Geologists, Mem. 17.

Hallam, A., and Wignall, P. 1997. Mass extinctions and their aftermath. Oxford, England: Oxford Univ. Press.

Hsu, K., and McKenzie, J. 1990. Carbon isotope anomalies at era boundaries: Global catastrophes and their ultimate cause. *Geol. Soc. Am. Special Paper* 247, pp. 61–70.

Isozaki, Y. 1994. Superanoxia across the Permo-Triassic boundary: Record in accreted deep-sea pelagic chert in Japan: In *Global environments and resources,* pp. 805–812. Canadian Society of Petroleum Geologists, Mem. 17.

Knoll, A.; Bambach, R.; Canfield, D.; and Grotzinger, J. 1996. Comparative earth history and Late Permian mass extinction. *Science* 273:452–457.

Marshall, C. 1990. Confidence intervals on stratigraphic ranges. *Paleobiology* 16:1–10.

References

Marshall, C., and Ward, P. 1996. Sudden and gradual molluscan extinctions in the latest Cretaceous of Western European Tethys. *Science* 274:1360–1363.

McLaren, D. 1970. Time, life and boundaries. *Journal of Paleontology* 44:801–815.

Morante, R. 1996. Permian and early Triassic isotopic records of carbon and strontium events in Australia and a scenario of events about the Permian–Triassic boundary. *Historical Geology* 11:289–310.

Pope, K.; Baines, A.; Ocampo, A.; and Ivanov, B. 1994. Impact winter and the Cretaceous-Tertiary extinctions: Results of a Chicxulub asteroid impact model. *Earth and Planetary Science Express* 128:719–725.

Rampino, M., and Caldeira, K. 1993. Major episodes of geologic change: Correlations, time structure and possible causes. *Earth Planetary Science Letters* 114:215–227.

Raup, D. 1979. Size of the Permo-Triassic bottleneck and its evolutionary implications. *Science* 206:217–218.

Raup, D. 1990. *Extinction: Bad genes or bad luck?* New York: Norton.

Raup, D. 1990. Impact as a general cause of extinction: A feasibility test. In *Global catastrophes in earth history*, ed. V. Sharpton and P. Ward, pp. 27–32. Geol. Soc. Am. Special Paper 247.

Raup, D. 1991. A kill curve for Phanerozoic marine species. *Paleobiology* 17:37–48.

Raup, D., and Sepkoski, J. 1984. Periodicity of extinction in the geologic past. *Proc. Nat. Acad. Sci.*, A81, p. 801–805.

Retallack, G. 1995. Permian–Triassic crisis on land. *Science* 267:77–80.

Schindewolf, O. 1963. Neokatastrophismus? Zeit. *Der Deutschen Geol. Gesell.* 114:430–445.

Schultz, P., and Gault, D. E. 1990. Prolonged global catastrophes from oblique impacts, in Sharpton, V. L. and Ward, P. D., eds., Global catastrophes in Earth history, An interdisciplinary conference on impacts, volcanism and mass mortality: Geological Society of America Special Paper 247, p. 239–261.

Sheehan, P.; Fastovsky, D.; Hoffman, G.; Berghaus, C.; and Gabriel, D. 1991. Sudden extinction of the dinosaurs: Latest Cretaceous, Upper Great Plains, U.S.A. *Science* 254:835–839.

Sigurdsson, H.; D'hondt, S.; and Carey, S. 1992. The impact of the Creta-ceous–Tertiary bolide on evaporite terrain and generation of major sul-furic acid aerosol. *Earth Planetary Science Letters* 109:543–559.

Stanley, S. 1987. *Extinctions.* New York: Freeman.

Stanley, S., and Yang, X. 1994. A double mass extinction at the end of the Pa-leozoic Era. *Science* 266:1340–1344.

Teichert, C. 1990. The end-Permian extinction. In *Global events in Earth history,* ed. E. Kauffman and O. Walliser, pp. 161–190.

Ward, P. 1990. The Cretaceous/Tertiary extinctions in the marine realm: A 1990 perspective. In *Geological Society of America Special Paper* 247, pp. 425–432.

Ward, P. 1994a. *The end of evolution.* New York: Bantam Doubleday Dell.

Ward, P. D. 1990. A review of Maastrichtian ammonite ranges. In *Geological Society of America Special Paper* 247, pp. 519–530.

Ward, P., and Kennedy, W. 1993. Maastrichtian ammonites from the Biscay region (France and Spain). *Journal of Paleontology, Memoir* 34, 67:58.

Ward, P.; Kennedy, W. J.; MacLeod, K.; and Mount, J. 1991. Ammonite and inoceramid bivalve extinction patterns in Cretaceous–Tertiary bound-ary sections of the Biscay Region (southwest France, northern Spain). *Geology* 19:1181.

Chapter 9. The Surprising Importance of Plate Tectonics

Armstrong, R. L. 1981. Radiogenic isotopes: The case for crustal recycling on a near-steady-state no-continental-growth Earth. *Philos. Trans. R. Soc. London Ser. A* 301:443–472.

Arrhenius, G. 1985. Constraints on early atmosphere from planetary accre-tion processes. *Lunar and Planetary Sciences Institute Rep* 85-01:4–7.

Beck, M. E., Jr. 1980. Paleomagnetic record of plate-margin tectonic pro-cesses along the western edge of North America. *J. Geophys. Res.* 85:7115–7131.

Broecker, W. 1985. *How to build a habitable planet.* Palisades, NY: Eldigio Press.

Card, K. D. 1986. Tectonic setting and evolution of Late Archean greenstone belts of Superior province, Canada. In *Tectonic evolution of greenstone belts,* ed. M. J. de Wit and L. D. Ashwal. *Lunar and Planetary Sciences Institute Tech. Rep.* 86-10:74–76.

Condie, K. C. 1984. *Plate tectonics and crustal evolution* 2d ed. Oxford, England: Pergamon Press.

Cox, A. 1973. *Plate tectonics and geomagmetic reversals*. San Francisco: Freeman.

Dalziel, I. W. D. 1992. On the organization of American plates in the Neoproterozoic and the breakout of Laurentia. *GSA Today* 2:237.

DePaolo, D. J. 1984. The mean life of continents: Estimates of continental recycling from Nd and Hf isotopic data and implications for mantle structure. *Geophys. Res. Lett.* 10:705–708.

Dietz, R. S. 1961. Continent and ocean basin evolution by spreading of the sea floor. *Nature* 190:854–857.

Goldsmith, D., and Owen, 1992. *The search for life in the universe*. Menlo Park, CA: Benjamin/Cummings.

Hess, H. H. 1962. History of ocean basins. In *Petrologic Studies—a volume to honor A.F. Buddington*, ed. A. E. J. Engel *et al.*, pp. 599–620. Boulder, CO: Geological Society of America.

Hoffman, P. F. 1988. United plates of America—the birth of a craton. *Ann. Rev. Earth Planet. Sci.* 16:543–603.

Howell, D. G., and Murray, R.W. 1986. A budget for continental growth and denudation. *Science* 233:446–449.

Hsü, K. J. 1981. Thin-skinned plate-tectonic model for collision-type orogenesis. *Sci. Sin.* 24:100–110.

Irving, E.; Monger, J. W. H.; and Yole, R. W. 1980. New paleomagnetic evidence for displaced terranes in British Columbia. In *The continental crust and its mineral deposits*, ed. D. W. Strangway. *Geol. Assoc. Canada Spec. Pap.* 20:441–456.

McElhinny, M. W. 1973. *Paleomagnetism and plate tectonics*. Cambridge, England: Cambridge Univ. Press.

Solomatov, V., and Moresi, L. 1997. Three regimes of mantle convection with non-Newtonian viscosity and stagnant lid convection on the terrestrial planets. *Geo. Res. Let.* 24:1907–1910.

Uyeda, S. 1987. *The new view of the earth*. San Francisco: Freeman.

Vine, F. J., and Mathews, D. H. 1963. Magnetic anomalies over oceanic ridges. *Nature* 199:947–949.

Wegener, A. 1924. *The origin of continents and oceans.* London: Methuen.

Wilson, J. T. 1965. A new class of faults and their bearing on continental drift. *Nature* 207:343.

Chapter 10. The Moon, Jupiter, and Life on Earth

Cameron, A. G. W. 1997. The origin of the moon and the single impact hypothesis V. *Icarus* 126:126–137.

Cameron, A. G. W., and R. M. Canup. 1998. The giant impact occurred during Earth accretion. *Lunar and Planetary Science Conference* 29: 1062.

Cameron, A. G. W., and R. M. Canup. 1999. State of the protoearth following the giant impact. *Lunar and Planetary Science Conference* 30:1150.

Chambers, J. E., and G. W. Wetherill. 1998. Making the terrestrial planets: N-body integrations of planetary embryos in three dimensions. *Icarus* 136:304–327.

Chambers, J. E.; Wetherill, G. W.; and Boss, A. P. 1996. The stability of multi-planet systems. *Icarus* 119:261–268.

Hartmann, W. K.; Phillips, R. J.; and Taylor, G. J. 1986. Origin of the moon. Lunar and Planetary Institute, 1986.

Wetherill, G. W. 1994. Possible consequences of absence of Jupiters in planetary systems. *Astrophys. and Space Sci.* 212:23–32.

Wetherill, G. W. 1995. Planetary science—how special is Jupiter? *Nature* 373:470.

Ida, S., and Lin, D. N. C. 1997. On the origin of massive eccentric planets: Detection and Study of Planets Outside the Solar System, 23rd meeting of the IAU, Joint Discussion 13, 25–26 August 1997, Kyoto, Japan. 13, E4

Chapter 11. Testing the Rare Earth Hypothesis

Beatty, J. K. 1996. Life from ancient Mars? *Sky and Telescope* 92:18.

Carr, M. H. 1998. Mars: Aquifers, oceans, and the prospects for life. *Astronomicheskii Vestnik* 32:453.

Chyba, C. F., *et al* 1999. Europa and Titan: Preliminary recommendations of the campaign science working group on prebiotic chemistry in the outer solar system. *Lunar and Planetary Science Conference* 30:1537.

Clark B. C. 1998. Surviving the limits to life at the surface of Mars. *J. Geophys. Res.* 103:28545.

Farmer J. 1998. Thermophiles, early biosphere evolution, and the origin of life on Earth: Implications for the exobiological exploration of Mars. *J. Geophys. Res.*, 103:28457.

Farmer, J. D. 1996. Exploring Mars for evidence of past or present life: Roles of robotic and human missions. *Astrobiology Workshop: Leadership in Astrobiology*, A59–A60.

Jakosky, B. M., and Shock, E. L. 1998. The biological potential of Mars, the early Earth, and Europa. *J. Geophys. Res.* 103:19359.

Kasting, J. F. 1996. Planetary atmosphere evolution: Do other habitable planets exist and can we detect them? *Astrophysics and Space Science* 241:3–24.

Klein, H. P. 1998. The search for life on Mars: What we learned from Viking. *J. Geophys. Res.* 103:28462.

Mancinelli, R. L. 1998. Prospects for the evolution of life on Mars: Viking 20 years later. Advances in Space Research 22:471–477.

McKay, C. P. 1996. The search for life on Mars. *Astrobiology Workshop: Leadership in Astrobiology, et al.* 12.

McKay, D. S., *et al.* 1996. Search for past life on Mars: Possible relic biogenic activity in Martian meteorite ALH84001. *Science* 273:924–930.

Nealson, K. H. 1997. The limits of life on Earth and searching for life on Mars. *J. Geophys. Res.*, 102:23675.

Owen, T., *et al.* 1997. The relevance of Titan and Cassini/Huygens to prebiotic chemistry and the origin of life on Earth. Huygens: Science, Payload and Mission, Proceedings of an ESA conference, ed. A. Wilson p. 231.

Shock, E. L. 1997. High-temperature life without photosynthesis as a model for Mars. *J. Geophys. Res.*, 102:23687.

Spangenburg, R., and D. Moser. 1987. Europa: The case for ice-bound life. *Space World* 8:284.

Chapter 12. Assessing the Odds

Caldeira K., and Kasting, J. 1992. The life span of the biosphere revisted. *Nature* 360:721–723.

Caldeira, K., and Kasting, J. F. 1992. Susceptibility of the early Earth to irreversible glaciation caused by carbon ice clouds. *Nature* 359:226–228.

Dole, S. 1964. *Habitable planets for man.* Waltham, MA: Blaisdell.

Gott, J. 1993. Implications of the Copernican Principle for our future prospects. *Nature* 363:315–319.

Gould, S. 1994. The evolution of life on Earth. *Scientific American* 271:85–91.

Hart, M. 1979. Habitable zones around main sequence stars. *Icarus* 33:23–39.

Kasting, J. 1996. Habitable zones around stars: An update. In *Circumstellar habitable zones*, ed. L. Doyle, pp. 17–28. Menlo Park, CA: Travis House.

Kasting, J.; Whitmire, D.; and Reynolds, R. 1993. Habitable zones around main sequence stars. *Icarus* 101:108–128.

Laskar, J.; Joutel, F.; and Robutel, P. 1993. Stabilization of the Earth's obliquity by the Moon. *Nature* 361:615–617.

Laskar, J., and Robutel, P. 1993. The chaotic obliquity of planets. *Nature* 361:608–614.

Lovelock, J. 1979. *Gaia, a new look at life on Earth.* Oxford, England: Oxford Univ. Press.

Marcy, G., et al. 1999. Planets around sun-like stars, Bioastronomy 99: A New Era in Bioastronomy. 6th Bioastronomy Meeting, Kohala Coast, Hawaii, August 2–6.

Marcy, G., and Butler, R. P. 1998. New worlds: The diversity of planetary systems. *Sky and Telescope* 95:30.

McKay, C. 1996. Time for intelligence on other planets. In *Circumstellar habitable zones*, ed. L. Doyle, pp. 405–419. Menlo Park, CA: Travis House.

Schwartzman, D., and Shore, S. 1996. Biotically mediated surface cooling and habitability for complex life. In *Circumstellar habitable zones*, ed. L. Doyle, pp. 421–443. Menlo Park, CA: Travis House.

Volk, T. 1998. *Gaia's body: Toward a physiology of Earth.* New York: Springer-Verlag.

Walker, J.; Hays, P.; and Kasting, J. 1981. A negative feedback mechanism for the long-term stabilization of Earth's surface temperature. *Journal of Geophysical Research* 86:9776–9782.

Chapter 13. Messengers from the Stars

Dick, S. 1982. *Plurality of worlds.* Cambridge, England: Cambridge Univ. Press.

Gleiser, M. 1997. *The dancing universe: From creation myths to the Big Bang.* New York: Dutton.

Gott, J. 1993. Implications of the Copernican Principle for our future prospects. *Nature* 363:315–319.

Wetherill, G. 1994. The plurality of habitable worlds. Fifth Exobiology Symposium and Mars Workshop, NASA Ames Research Center.

Index

317